THE ART OF WOODWORKING

FINISH CARPENTRY

THE ART OF WOODWORKING

FINISH CARPENTRY

TIME-LIFE BOOKS
ALEXANDRIA, VIRGINIA

ST. REMY PRESS
MONTREAL • NEW YORK

THE ART OF WOODWORKING was produced by
ST. REMY PRESS

PUBLISHER	Kenneth Winchester
PRESIDENT	Pierre Léveillé
Series Editor	Pierre Home-Douglas
Series Art Director	Francine Lemieux
Senior Editor	Marc Cassini
Editor	Jim McRae
Art Directors	Normand Boudreault, Luc Germain, Solange Laberge
Designers	Hélène Dion, Michel Giguère
Picture Editor	Christopher Jackson
Writers	Andrew Jones, Rob Lutes, David Simon
Research Assistant	Bryan Quinn
Contributing Illustrators	Gilles Beauchemin, Roland Bergerat, Michel Blais, Jean-Guy Doiron, Ronald Durepos, Robert Paquet, Maryo Proulx, James Thérien
Administrator	Natalie Watanabe
Production Manager	Michelle Turbide
System Coordinator	Jean-Luc Roy
Photographer	Robert Chartier
Administrative Assistant	Dominique Gagné
Proofreader	Garet Markvoort
Indexer	Christine M. Jacobs

Time-Life Books is a division of Time Life Inc.,
a wholly owned subsidiary of
THE TIME INC. BOOK COMPANY

TIME-LIFE INC.

President and CEO	John M. Fahey
Editor-in-chief	John L. Papanek

TIME-LIFE BOOKS

President	John D. Hall
Vice-President, Director of Marketing	Nancy K. Jones
Executive Editor	Roberta Conlan
Executive Art Director	Ellen Robling
Consulting Editor	John R. Sullivan
Production Manager	Marlene Zack

THE CONSULTANTS

Karl Marcuse is a self-employed carpenter and contractor in Montreal. He has worked as a home renovator in many countries and is now completing restoration of his century-old home.

Giles Miller-Mead taught advanced cabinetmaking at Montreal technical schools for more than ten years. A native of New Zealand, he has worked as a restorer of antique furniture.

Scott Schuttner is a carpenter and home builder in Fairbanks, Alaska. A frequent contributor to *Fine Homebuilding*, he has been building staircases and teaching aspiring carpenters for close to 20 years.

Joseph Truini is Senior Editor of *Home Mechanix* Magazine. A former Shop and Tools Editor of *Popular Mechanics*, he has worked as a cabinetmaker, home improvement contractor and carpenter.

Finish carpentry
p. cm. — (The Art of woodworking)
Includes index.
ISBN 0-8094-9520-1
1. Finish carpentry I. Time-Life Books. II. Series.
TH5640.F563 1994
694'.6—dc20 94-13533
CIP

For information about any Time-Life book,
please call 1-800-621-7026, or write:
Reader Information
Time-Life Customer Service
P.O. Box C-32068
Richmond, Virginia
23261-2068

© 1994 Time-Life Books Inc.
All rights reserved.
No part of this book may be reproduced in any form or by any electronic or mechanical means, including information storage and retrieval devices or systems, without prior written permission from the publisher, except that brief passages may be quoted for reviews.
First printing. Printed in U.S.A.
Published simultaneously in Canada.

TIME-LIFE is a trademark of Time Warner Inc. U.S.A.

CONTENTS

6 INTRODUCTION

12 FINISH CARPENTRY BASICS
14 Finish carpentry tools
16 Basic cuts

20 MOLDING
22 Molding styles
23 Baseboard
30 Chair and picture rails
33 Crown molding

38 PANELING
40 Paneling styles
42 Tongue-and-groove wainscoting
46 Frame-and-panel wainscoting
52 Paneled ceilings

56 WINDOWS
58 Basic window trim styles
59 Installing windows
61 Picture-frame casing
69 Stool-and-apron casing
75 Making a window sash
81 A glazing bar half-lap joint

84 DOORS
86 Anatomy of a door
88 Tools and door hardware
90 Frame-and-panel doors
95 Door jambs
100 Hanging a door
107 Locksets

114 STAIRS
116 Anatomy of a staircase
118 The stringers
123 Treads and risers
128 Newel posts
132 Handrails
136 Balusters

140 GLOSSARY

142 INDEX

144 ACKNOWLEDGMENTS

Grant Taylor describes

HOW I BUILD DOORS

I make custom, high-quality hardwood doors for a living, but no matter how many doors I see, I always marvel at the craftmanship involved: The way a door maker can bring out the wood grain by using quartersawn boards, for example, or how the mortise-and-tenon joinery makes a joint that won't open up for 100 years or more. But there's something else besides the craftmanship that always strikes me. It doesn't matter whether it's a thick oak door on a medieval English castle or a modern stained-glass assembly with a delicate arching sash. In some way, all doors are magic, offering us the possibilty of mystery or the unexpected, just beyond the turn of the knob.

I build my doors much as door builders of old worked. For starters, I use local wood that has been cut and milled by woodmen I know. When the felled trees are lying in fresh stacks, I climb over the logs and select prime pieces for milling. Those rough boards are later dropped off at my shop—a stone structure that I built myself in rural New Hampshire—where I carefully mill them to reveal their unique grain patterns. I select the finest specimens and then dry them in a solar kiln that I also constructed. After proper aging and drying—a process that gives optimum stability to the wood—I finally bring into my shop a piece of wood that probably has been touched by only a couple of people since it stood as a tree in the forest.

Cherry and oak are my favorite woods, and they grow wonderfully strong in the area where I live. Their grain patterns are invariably spectacular, and no matter how many times I assemble a door—typically I work from custom design plans, so every one is unique—there is always a thrill when I pull the milled boards out of the planer and marvel at the pattern that is revealed.

Though I use many traditional hand tools to assemble my doors, I rely on power tools to get the precision my clients come to expect: Tolerances of 1/64 inch in door pieces such as stiles, rails, and panels are common in my shop.

I am proud of what I produce, and it's never boring. The range of styles that people look for in custom doors always keeps me on my toes, always doing something interesting. Whether I'm working with a local blacksmith to fabricate some wrought-iron hinges for a Tudor-style door or figuring out the complicated geometry of cutting center ovals in a door that's taken me a week to complete, I never have a dull day. The only thing I find disagreeable is when the process comes to an end. These doors are something I have poured my heart into, something I've sweated over to make beautiful. I just hate to see them go.

Grant Taylor is the owner of Lamson-Taylor Custom Doors, a two-person shop in South Acworth, New Hampshire. Taylor has built custom doors for houses throughout North America.

Jon Eakes on

MAKING TRIM AND MOLDING

I walked in and I knew within two minutes that this was a quality-built house. It was a modest split-level—no vaulted ceilings, no spectacular centerpiece. In fact there was nothing "outstanding." Even the trim was simple and unobtrusive, but a close look around the windows showed precise mitered angles and no sign of nails. It was the trim around the bottom of the wall that said the most: tight corners and carefully constructed returns. Very few people bother to do that today.

Trim and moldings were once the most obvious part of both furniture and house interior finishing. There were simple mechanical reasons why most of it has disappeared today. The beauty and creativity of molding designs was an outgrowth of the need to hide construction joints as well as junctions between different materials. Modern materials have changed all that. With drywall returns on windows and tapered corners, many modern houses now use trim only to hide the door frames and the intersection between walls and floors.

We can bemoan the sterilized look of particleboard furniture and box-like houses—or we can see the absence of joints as liberating molding from its mechanical need to hide something, allowing its shape and placement to be determined solely by our esthetic desires.

Making your own moldings is one of the most satisfying ways of letting your creativity show in your woodworking. Although routers can easily decorate edges and with a bit of work even make full moldings, it is the table saw outfitted with a three-blade molding head that can really produce. In my experience the keys to success in making molding or trim on a table saw are very simple:

- Use very sharp knives.
- Use wood that is either flat or easily pressed flat on the table.
- Use firm hold-downs that prevent vibration.
- Advance the wood just fast enough to prevent burning but just slowly enough to avoid "waves" on the face of the wood.

Honing your molding knives for that very clean cut is easier than it may seem. Never try sharpening the curved end: You won't get all three to match. Lay each knife flat on a very flat sharpening stone and grind the entire side of the knife, the same length of time for each one. You won't make it much thinner but you will hone the cutting edge and keep all three blades exactly the same shape and length.

Jon Eakes has been a cabinetmaker and custom renovator in Montreal, Canada for more than 20 years. He is known primarily for his teaching through books, videos, radio, and the TV show Renovation Zone.

Scott Schuttner on the CHALLENGE OF BUILDING STAIRS

I learned stairbuilding the way most of us do—from books and trial and error. However, the books of 30 years ago usually covered such subjects either with rudimentary abruptness or arcane terminology. Neither approach satisfied me and I have come to understand that there is more to consider than the obvious function of delivering people from one elevation to another.

A graceful staircase is one of the most prominent architectural features in a home and a showcase of a woodworker's talents. The care and patience that go into building stairs may be on display for 100 years and not easily remodeled or repaired. A staircase must be made rock solid, as it will serve as a playground, slide, and raceway for children.

A stairbuilder must be aware that users become physically attuned to the stairs. Our legs quickly pick up a rhythm for a set of stairs, which after one or two steps allows us to negotiate the rest without all-absorbing attention.

When faced with these concerns, it is not surprising that novice carpenters shy away from building a stairway. A minor miscalculation can result in one step being out of sync with the rest and even if not discernable visually, it may result in a stumble for the unwary.

But not to worry. Although stairbuilding requires thorough planning and attention to detail, in most cases the math is straightforward and the carpentry, although challenging, need not be intimidating.

To help ensure success I always draw a precisely dimensioned, side-view sketch of the stairs and include all details such as tread thickness, floor coverings, landings, and rough framing while paying close attention to the first and last step since this is where most mistakes are likely to occur. And initially, I plan vertical distances with reference to *finished* floor and tread surfaces and make adjustments from there.

I also keep in mind building codes that dictate limits for the rise and run, width of stairs, sizes of handrails, and headroom clearances to name a few. Codes do not necessarily prescribe great stairs but they will keep you from building disastrous ones.

As a stairbuilder in Alaska, I am asked to use a wide variety of materials combined into many eclectic styles, ranging from rustic treads and carriages using split spruce logs to grand staircases with mitered nosings, volutes, and goosenecks. While no two ever look the same, their construction follows the same processes. It's always a lot of fun and the end result turns out to be useful for so much more than merely enabling people to go upstairs or down.

Scott Schuttner is the author of Basic Stairbuilding and other books and articles by The Taunton Press. He lives and operates his company, Heartwood Builders and Woodworking, in Fairbanks, Alaska.

FINISH CARPENTRY BASICS

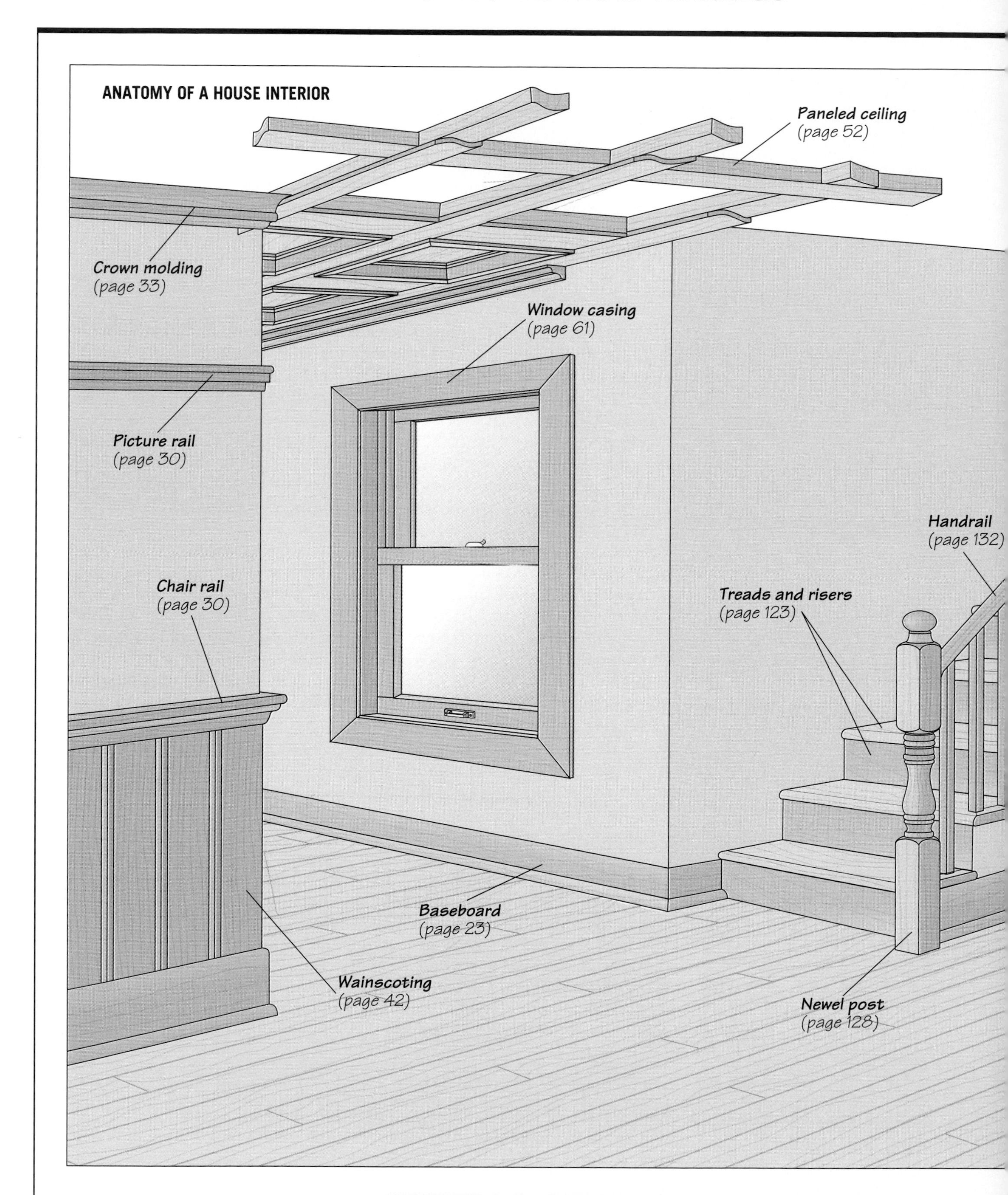

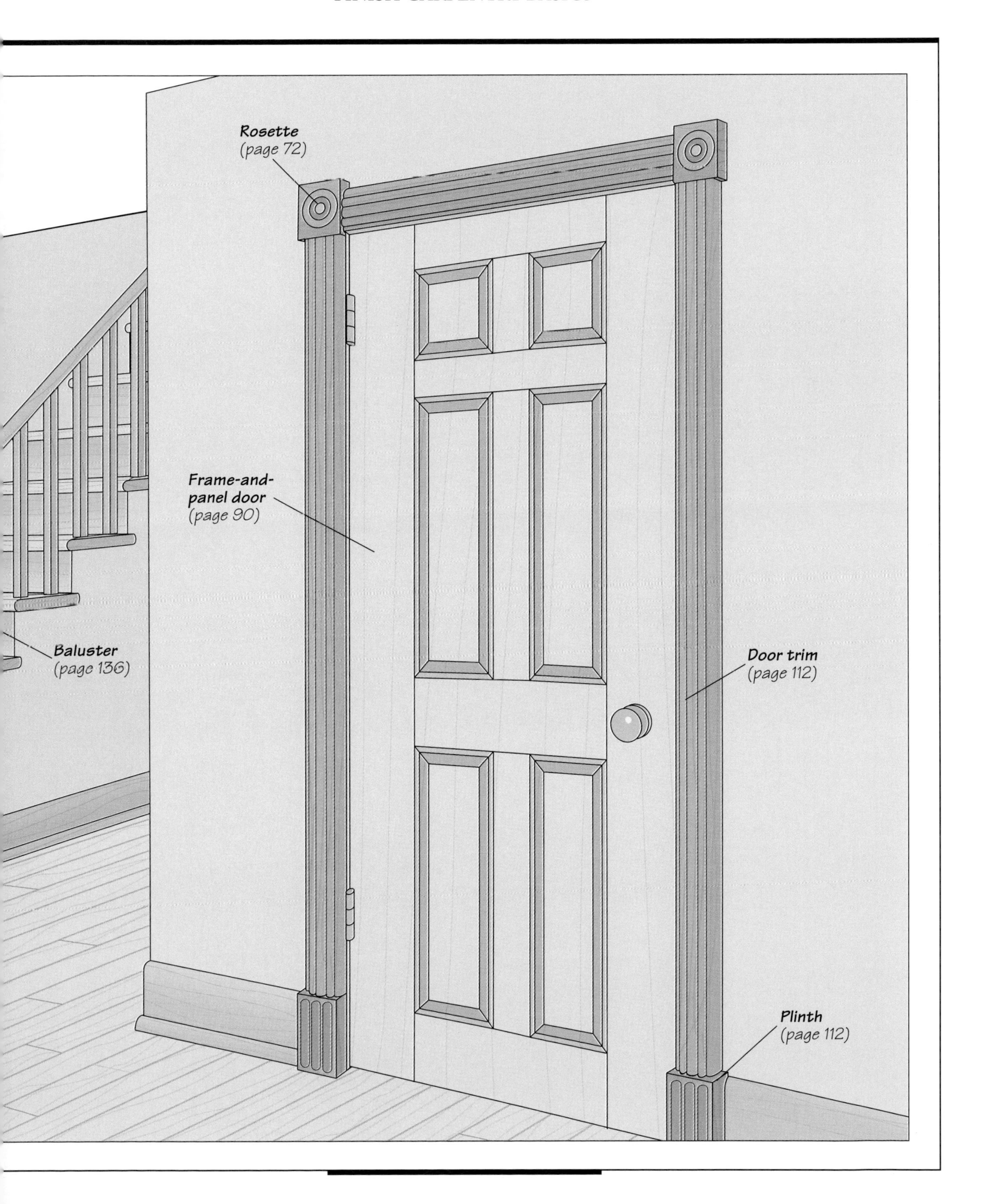
Rosette
(page 72)
Frame-and-panel door
(page 90)
Baluster
(page 136)
Door trim
(page 112)
Plinth
(page 112)

FINISH CARPENTRY TOOLS

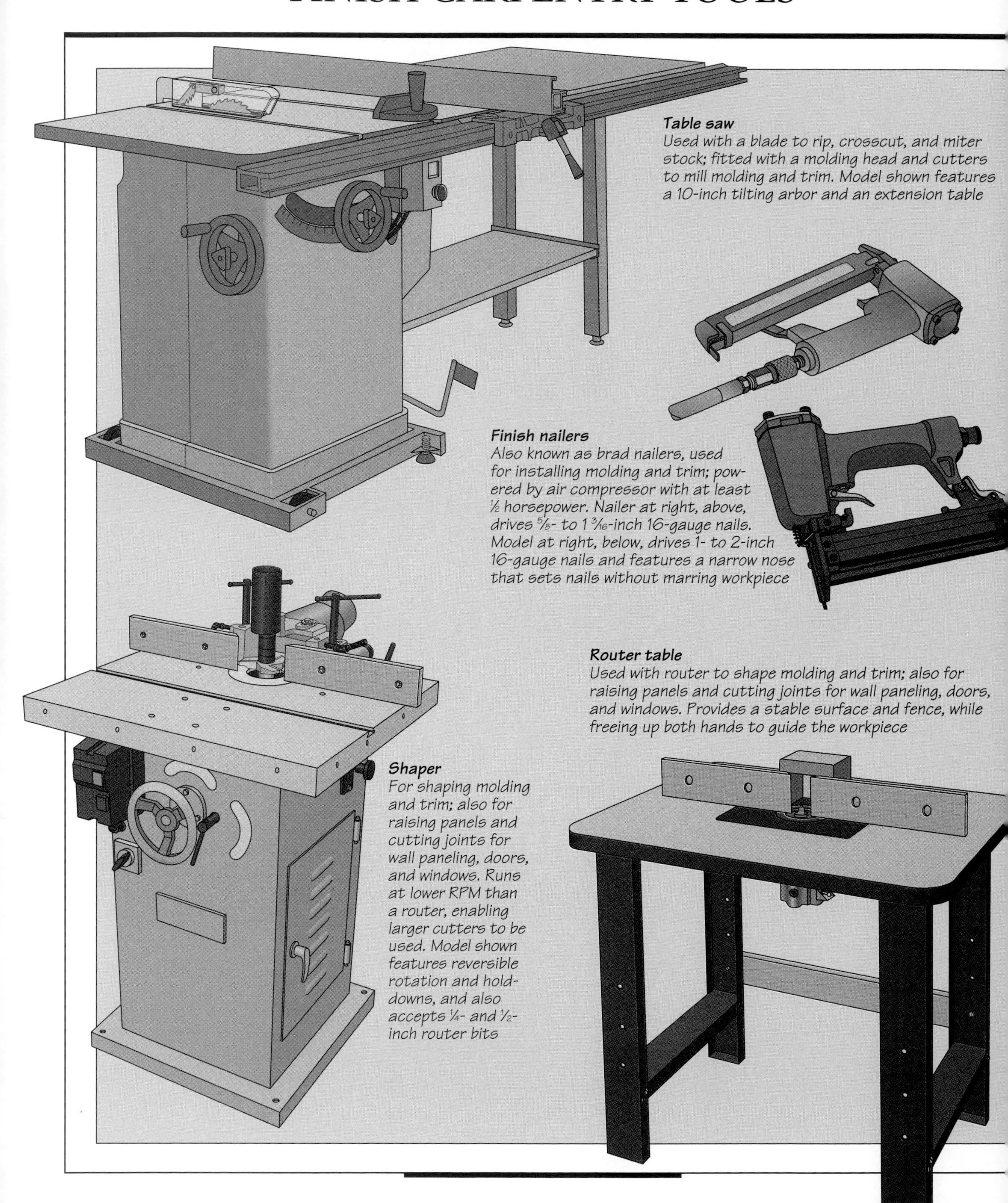

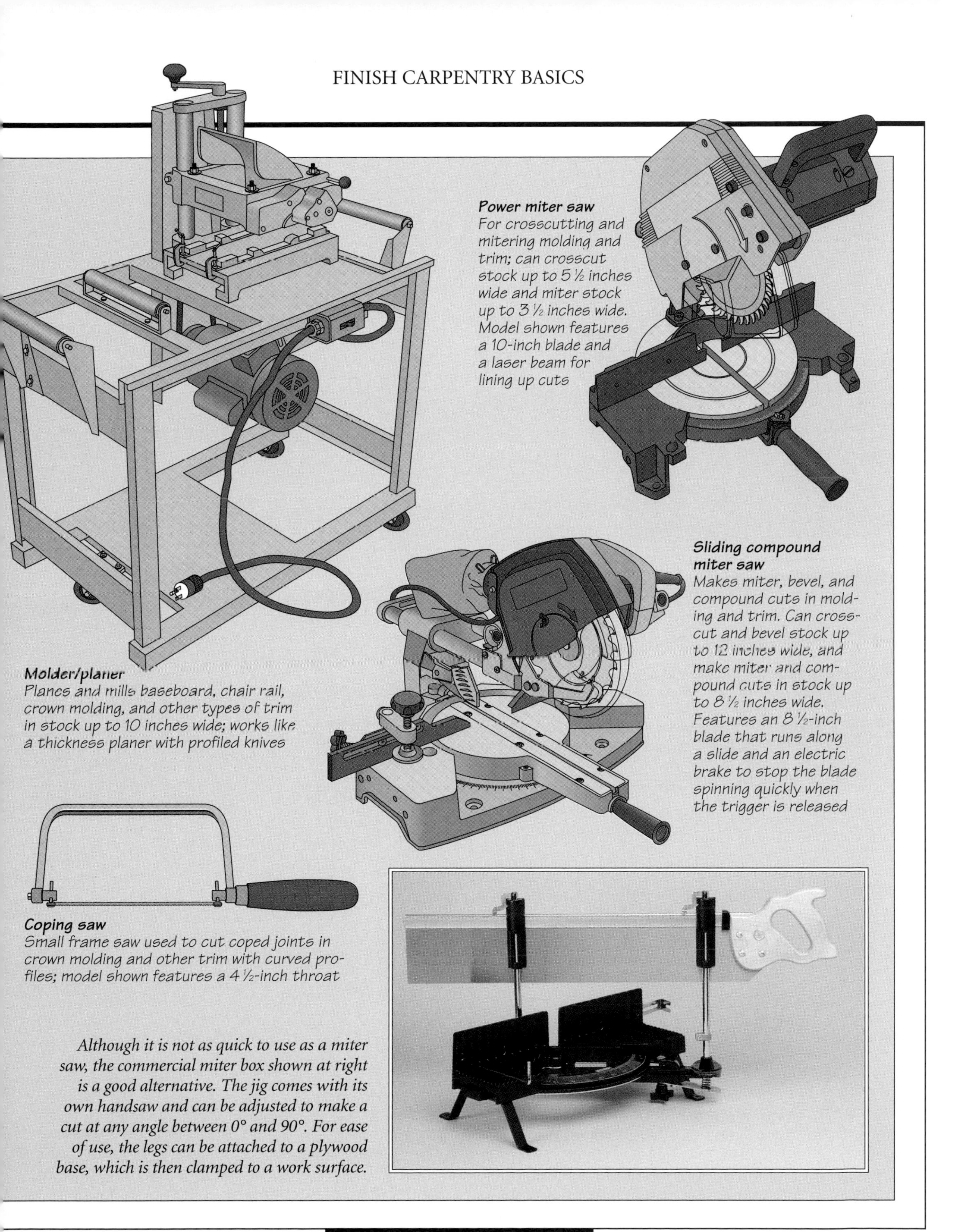

Although it is not as quick to use as a miter saw, the commercial miter box shown at right is a good alternative. The jig comes with its own handsaw and can be adjusted to make a cut at any angle between 0° and 90°. For ease of use, the legs can be attached to a plywood base, which is then clamped to a work surface.

A sliding compound saw is set up to miter a length of molding. It is a good idea to mount the saw on a portable miter stand, which enables you to work at a comfortable height. The model shown features support arms that can be adjusted to extend 4 feet on each side of the blade to accommodate long workpieces.

PREPARING STOCK

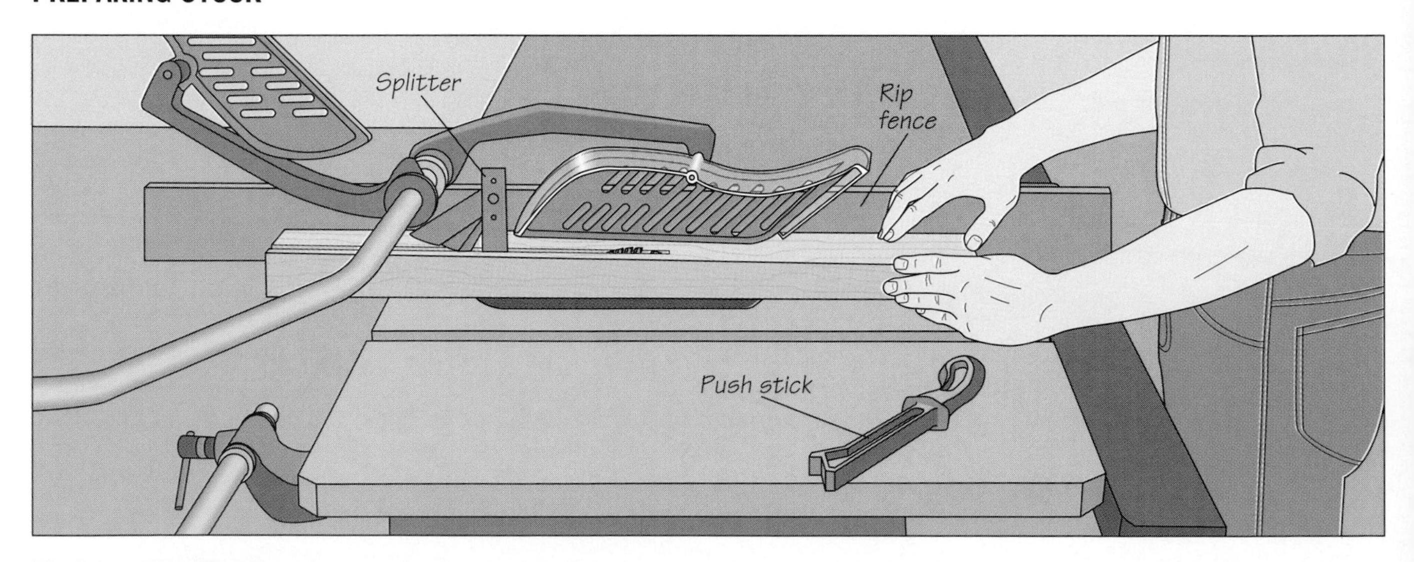

Ripping on the table saw

If you are using rough lumber for your finish carpentry projects, start by jointing one face of each board, and then an edge. Next, rip the board to width, making the second edge parallel to the jointed edge. Set the stock face down on the saw table and adjust the blade height about ¼ inch above the workpiece. Position the rip fence for the width of the cut, then feed the stock into the blade, holding it firmly against the fence *(above.)* Stand slightly to the left of the workpiece and straddle the fence with your right hand, making certain that neither hand is in line with the blade. Once your fingers approach the blade guard, use a push stick to complete the cut. **(Caution: Blade guard partially retracted for clarity.)**

Crosscutting on the table saw
To cut the board to length, hold it flush against the miter gauge, and align your cutting mark with the blade. Position the rip fence well away from the end of the stock to prevent the cut-off piece from jamming against the blade and kicking back. Hook the thumbs of both hands over the miter gauge to hold the stock firmly against the gauge and flat against the table, then feed the board into the blade *(right)*. **(Caution: Blade guard partially retracted for clarity.)**

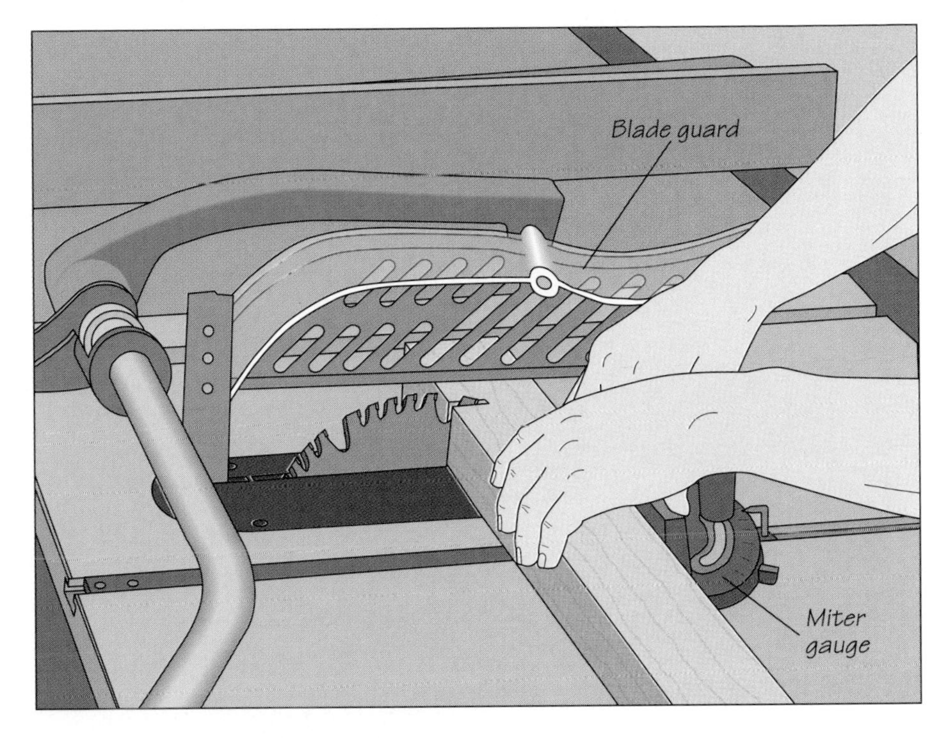

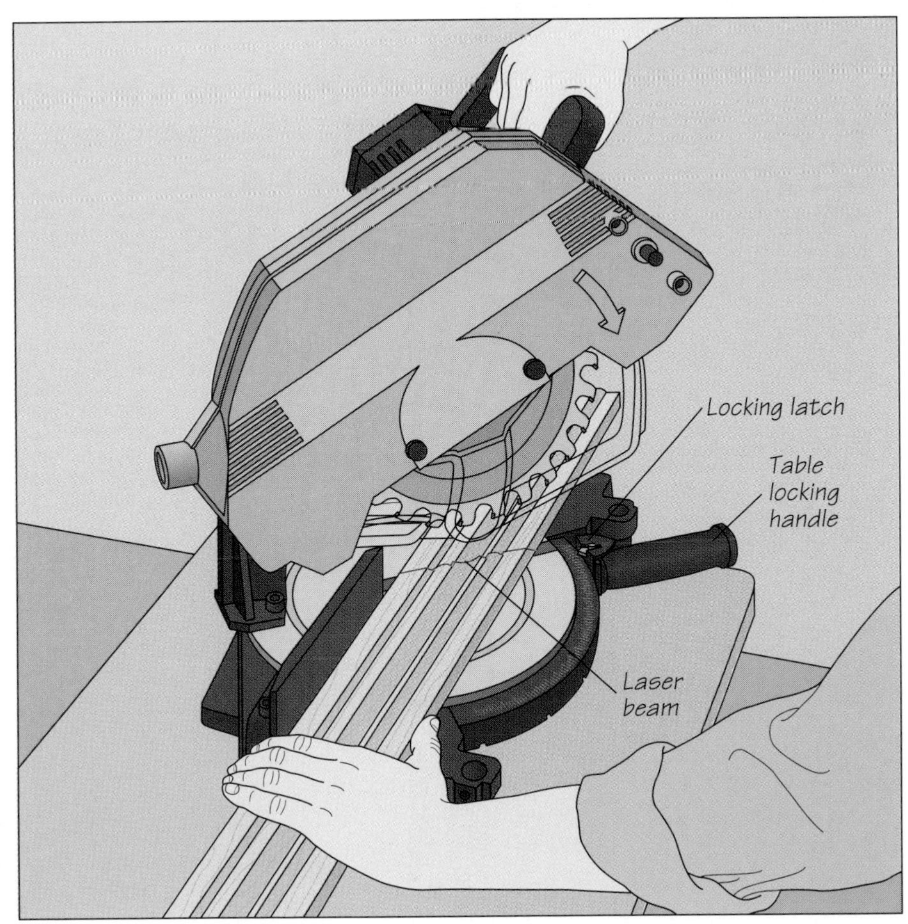

Making a miter cut
Adjust the saw to the desired miter angle. On the model shown at left, turn the table locking handle counterclockwise, depress the locking latch, and swing the table left or right until the pointer indicates the appropriate angle. Turn the handle clockwise to lock the table. Set your workpiece on the table and align the cutting mark with the table slot. The model shown features a laser beam to help you line up the cutting mark. Holding the workpiece firmly against the table and fence, turn on the saw by squeezing the handle trigger and bring the saw down slowly *(left)*. Once the cut is completed, release the trigger and lift the handle until the blade clears the workpiece.

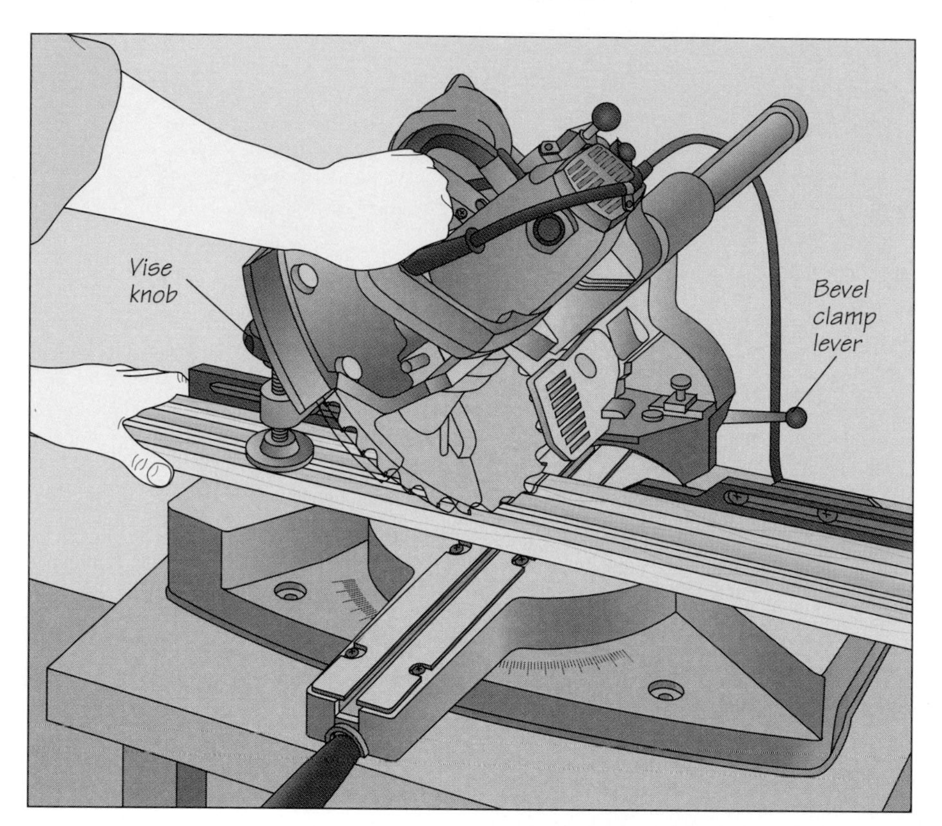

Making a bevel cut on a sliding compound saw
Adjust the saw to the desired bevel angle. On the model shown, loosen the bevel clamp lever, tilt the blade assembly to the left, and set the bevel to the required angle. Tighten the clamp lever. Set the workpiece against the fence, aligning the cutting line with the blade, and secure it in place using the vise knob. To make the cut, grip the handle and slide the blade assembly forward. Squeeze the trigger in the handle, bring the handle down, and slide the saw blade back to cut the workpiece *(left)*.

Making a compound cut on a sliding compound saw
Adjust the saw to the desired bevel and miter angles. On the model shown, start by setting the bevel angle *(above)*. To set the miter angle, loosen the table locking handle and swing the table to the left or right to the desired angle. Set the workpiece against the fence, aligning your cutting line just to the waste side of the blade. Clamp the workpiece in place using the vise knob. Make the compound cut *(right)* as you would a bevel cut.

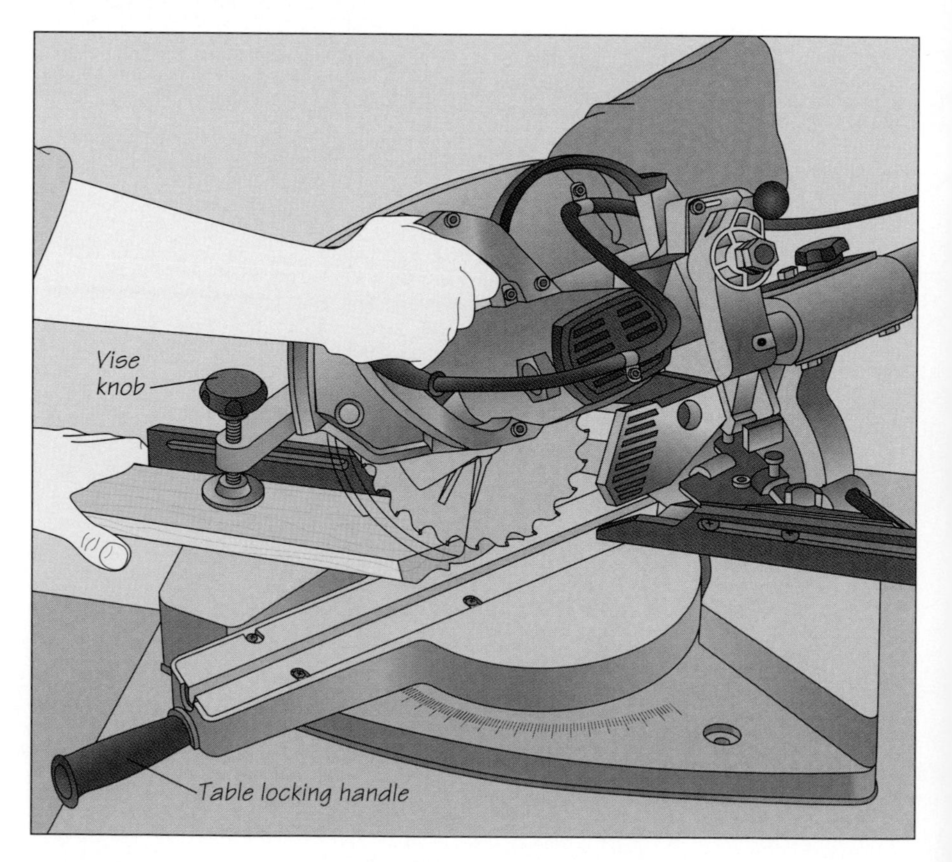

WOOD SPECIES FOR FINISH CARPENTRY

WOOD SPECIES	STRENGTH	WORKABILITY	GLUING QUALITY	FINISHING
Ash	Fair	Fair	Fair	Accepts stains well; requires heavy filler for painting
Basswood	Poor	Good	Good	Accepts stains well
Beech	Fair	Fair	Fair	Accepts stains well; requires thin filler for painting
Birch	Good	Good	Fair	Accepts stains well; requires thin filler for painting
Cedar, Western red	Poor	Good	Good	Oil stain recommended
Cherry	Fair	Good	Fair	Accepts stains well; not suitable for painting
Cypress	Fair	Good	Fair	Oil stain recommended
Douglas-fir	Fair	Good	Good	Oil stain recommended
Elm	Good	Good	Fair	Accepts stains well; requires heavy filler for painting
Gum, sweet red	Fair	Good	Good	Accepts stains well; requires thin filler for painting
Hemlock	Fair	Fair	Good	Oil stain recommended
Hickory	Good	Difficult	Fair	Accepts stains well; not suitable for painting
Mahogany	Fair	Good	Good	Accepts stains well; not suitable for painting
Maple, hard	Good	Difficult	Fair	Accepts stains well; requires thin filler for painting
Maple, soft	Fair	Fair	Fair	Accepts stains well; requires thin filler for painting
Oak, red	Good	Good	Fair	Accepts stains well; requires heavy filler for painting
Oak, white	Good	Good	Fair	Accepts stains well; requires heavy filler for painting
Pine, ponderosa	Poor	Good	Good	Accepts finishes well
Pine, yellow	Poor	Fair	Fair	Accepts finishes well
Redwood	Fair	Good	Good	Oil stain recommended
Spruce	Poor	Good	Good	Accepts finishes well
Teak	Good	Difficult	Poor	Oil stain recommended; not suitable for painting
Walnut	Good	Good	Good	Accepts stains well; not suitable for painting

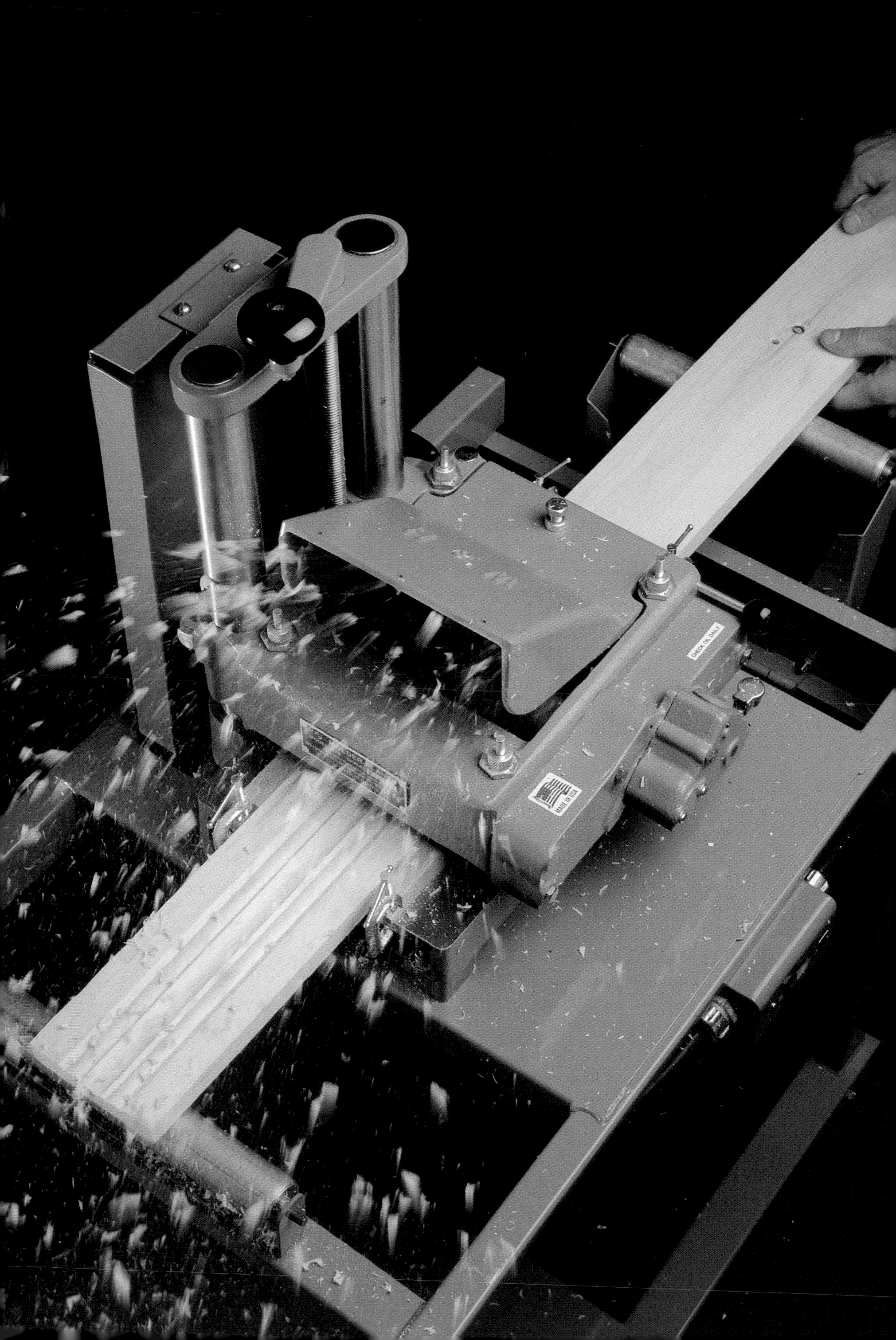
CHECK OIL DAILY
MADE IN USA

MOLDING

For woodworkers with a lot of molding to install, the finish nailer is a handy alternative to a hammer. Here, the nailer secures a length of crown molding to the ceiling. Powered by compressed air, the nailer drives home small-gauge finishing nails without splitting the wood.

To the Greeks and Romans, proportion was in the design of everything they built. For visual appeal, their structures relied on a logical and harmonious progression of architectural elements, one atop the other, from plinth to cornice. Some 2000 years later, furniture makers of the Georgian period used small-scale versions of the same elements to decorate the interiors of their patrons' homes.

Today, molding is a broad term that encompasses all interior trim applied to walls and ceilings, such as baseboard, chair rail, picture rail, and crown molding. An example of each type is illustrated in a typical house interior on page 12. This chapter presents instructions for making and installing these different kinds of molding.

Whether it is the angular trim of an Arts and Crafts-style home or the formal cornice of a Victorian parlor, molding serves a functional as well as a decorative role. Baseboard *(page 23)*, for example, is designed to cover gaps between the wall and the floor, while crown molding serves the same purpose along the ceiling. Chair rails *(page 30)* prevent chair backs from nicking walls and paneling, and picture rails provide a handy way to hang art without marring walls.

The advent of the molding machine in the 19th Century made it possible to mass-produce this functional and decorative material. Today, you can buy the most popular profiles of crown molding and baseboard at virtually any hardware store. Specialized millwork shops stock a wider range of profiles, and some will custom-grind special knives so that an antique pattern can be reproduced. But molding is also easy to make in the shop *(page 24)*. All you need is a table saw with a tilting arbor and a molding head or a table-mounted ½-inch router—and a bit of imagination. If you plan to produce a great deal of molding, a shaper or a molder/planer like the one shown in the photo at left may be a worthwhile purchase.

Installing molding can be a simple task once you have mastered a few basic principles *(page 26)*. Use longer pieces for the main rooms so that there will be fewer joints in these locations. Save shorter pieces for inside closets and less conspicuous areas of the home.

Combining features of the thickness planer and the shaper, a molder/planer mills a length of chair rail. The machine works much like a planer, except that it can be fitted with custom-ground knives that match the desired profile. The model shown at left has the power and capacity to turn out custom baseboards, chair rails, and crown molding quickly and accurately.

MOLDING STYLES

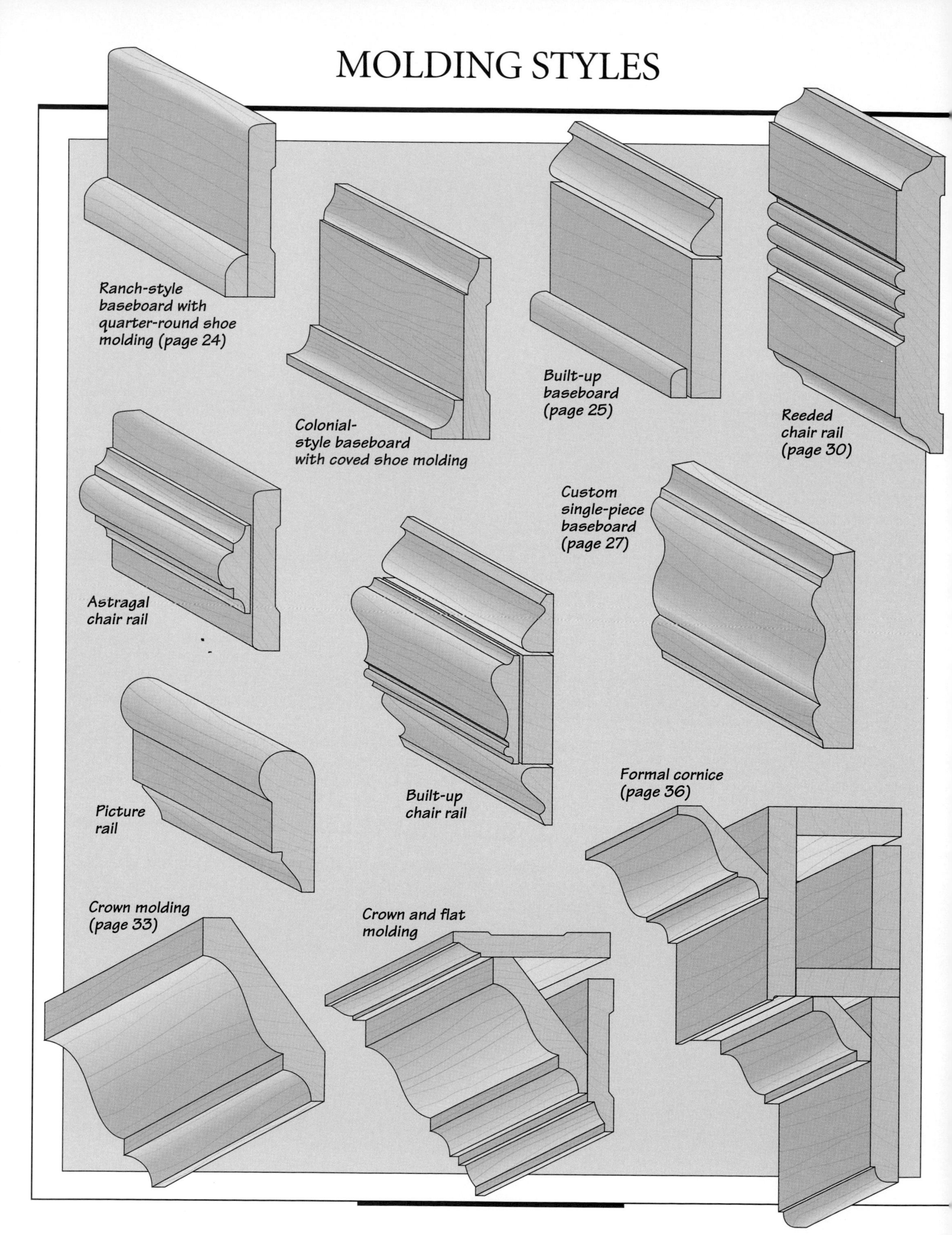

BASEBOARD

Also known as skirt, baseplate, mopboard, or just plain base, baseboard is the most common form of molded running trim used in finish carpentry. Baseboards serve a dual purpose: they visually anchor the wall to the floor and they also cover any gaps between those two surfaces.

Baseboard comes in two basic types: single-piece or built-up. Standard single-piece baseboard is usually between 3 and 12 inches wide and is sold in a variety of molded profiles; it can be made easily with a table saw, router, or shaper. While single-piece baseboard may be easier to install, it is more likely to cup than built-up molding. The simplest form of built-up baseboard is base-and-shoe *(page 24)*, which features a molded shoe that provides a visual transition between the wall and floor.

Baseboard is installed with a hammer, a nail set and finishing nails, or with an air-powered finish nailer *(page 21)*. While the nailer is a more expensive alternative, it makes installation quick and clean. It is preferable to hand-nailing when working with hardwood molding, which is more prone to splitting than softwood. Baseboard molding is typically nailed in place after the walls have been painted and the finish floor installed and sanded. Then the shoe molding is nailed in place. If the floor is to be installed after the baseboard, leave a space under the baseboard for the floor; use scraps of the flooring to help you determine the size of the gap. As with all molding, baseboard can be stained or painted; this is best done after all the molding has been cut to size, but before it is installed.

The last piece of baseboard to be installed in a room often ends at a door casing. Here a U-shaped, shop-made jig called a preacher is used to measure the piece before it is nailed in place. The jig slips over the baseboard and rests against the casing, allowing you to mark the baseboard to length with precision.

MOLDING JOINERY

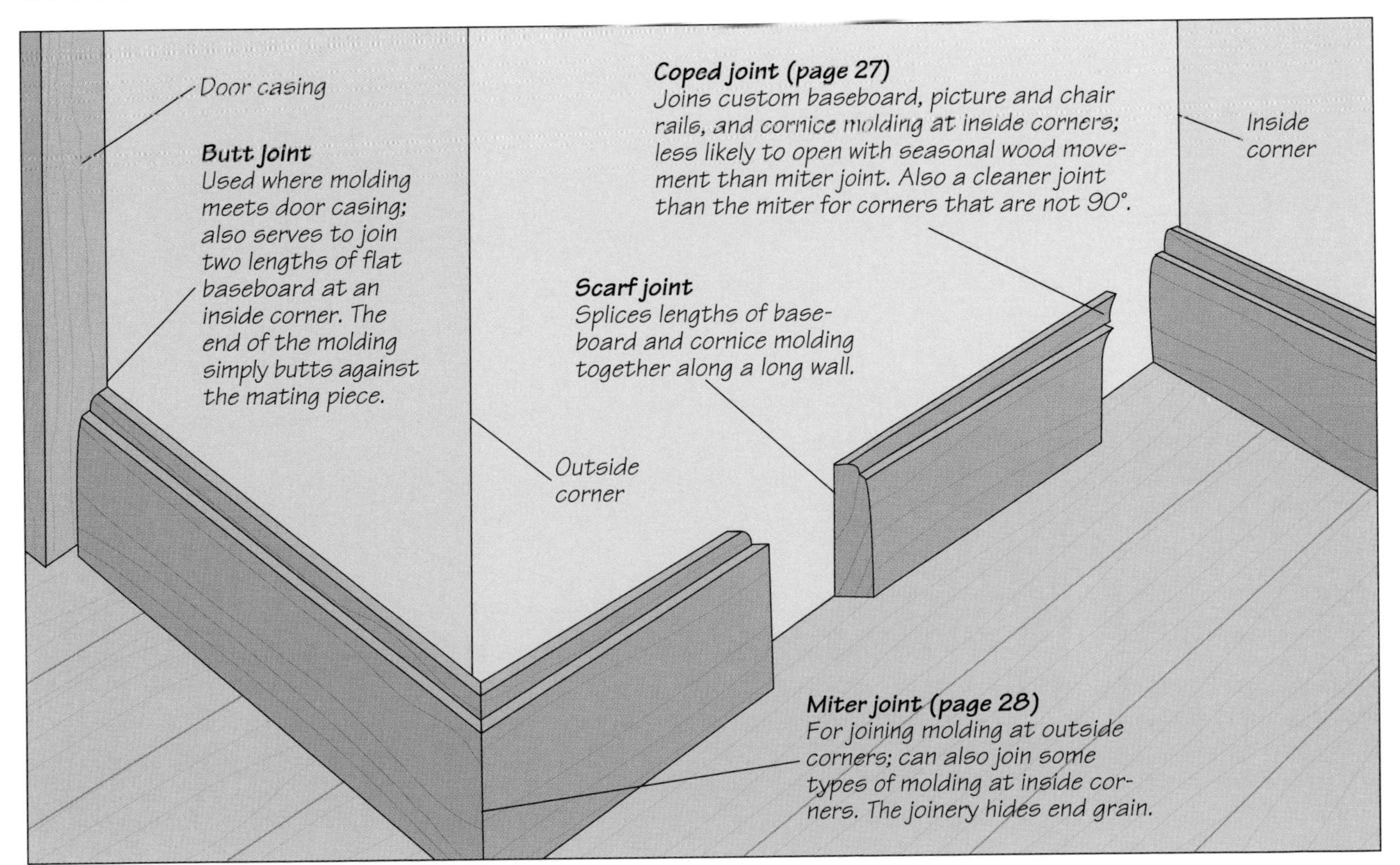

MAKING BASE-AND-SHOE BASEBOARD

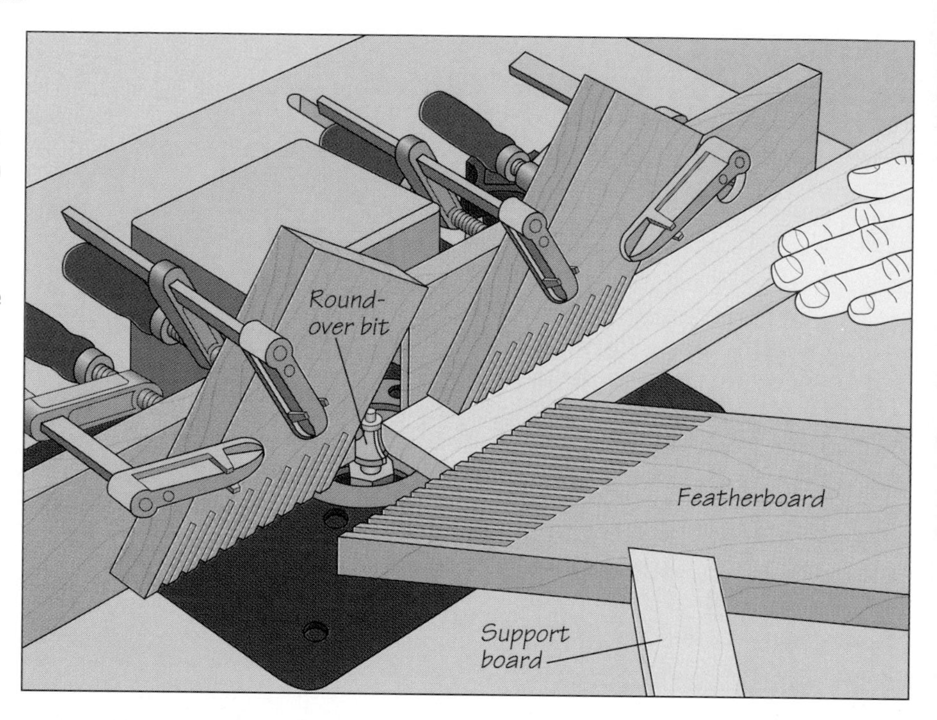

1 Milling the shoe on a router table
Make the shoe molding portion of the baseboard from ½-inch-thick stock. Install an edge-forming bit in your router and mount the tool in a table; a round-over bit is shown at right, but any other shape can be used. To support the workpiece, use three featherboards: Clamp two to the fence—one on each side of the bit—and a third to the table in line with the cutter. Shape both edges of each workpiece, feeding the stock with both hands and finishing the pass with a push stick. For safety, it is best to shape long boards that are at least 4 inches wide, and then rip the shaped edges off on the table saw.

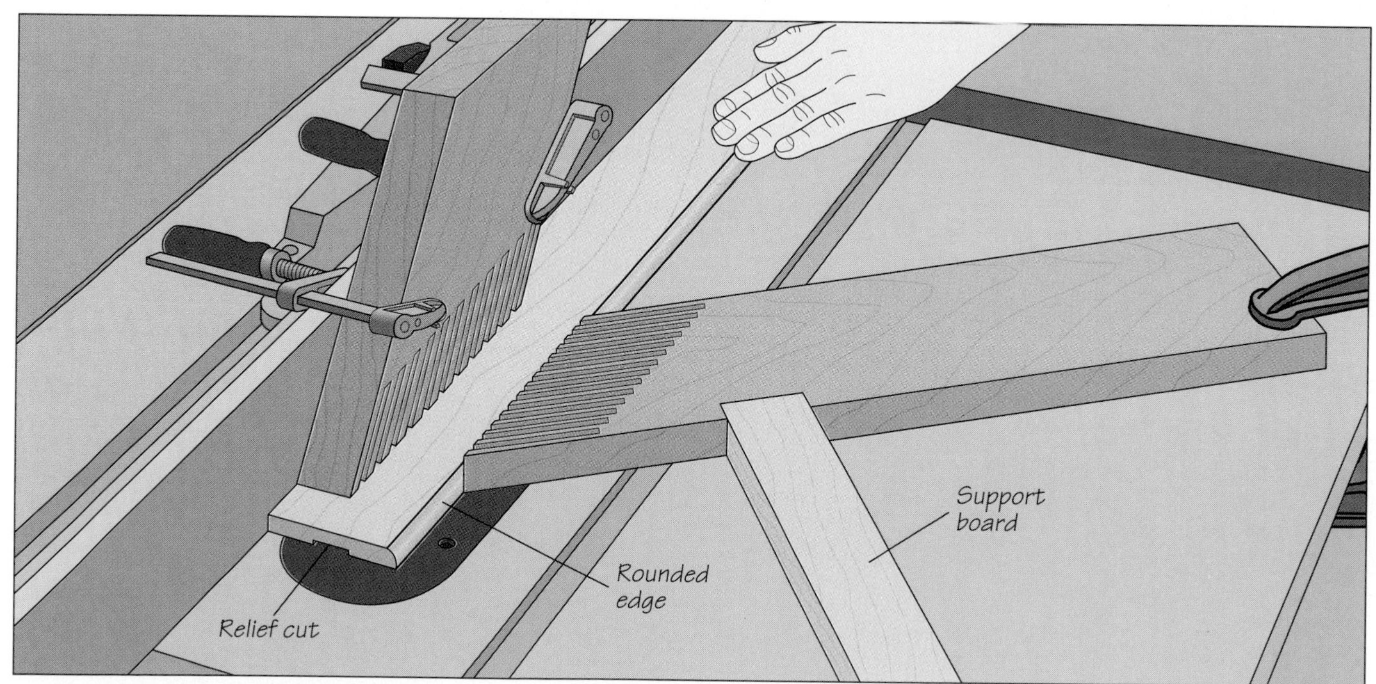

2 Relieving the base on the table saw
To prepare the base portion of the baseboard, shape one edge of a 1-by-4 or 1-by-6 as in step 1. Then plane the stock to the desired thickness. To prevent the molding from cupping and compensate for any irregularities in the wall, make a relief cut along the back face of the stock. Install a dado head on your table saw, adjusting its width to about 2 inches, and set the cutting height at ⅛ inch. Position the rip fence so the cut will be centered in the middle of the workpiece. Use two featherboards to support the stock and both hands to feed it face up, while butting it against the fence *(above)*. Finish the pass with a push stick.

MAKING BUILT-UP BASEBOARD

Milling a base cap

To make the top piece—the base cap—of the built-up baseboard illustrated on page 22, install a decorative edging bit on your shaper. (A router can do the job too, but the shaper enables you to employ larger cutters—and therefore thicker stock. It is also a more stable tool to use.) Keep the stock pressed against the fence and the table using the hold-downs supplied with the machine. Feed the stock on edge, using both hands *(right)*. To finish the pass, move to the outfeed side of the table and pull the stock past the cutter. The bottom piece of built-up baseboard is installed the same way as single-piece molding *(page 26)*. The base cap is then nailed to both the wall and the bottom piece.

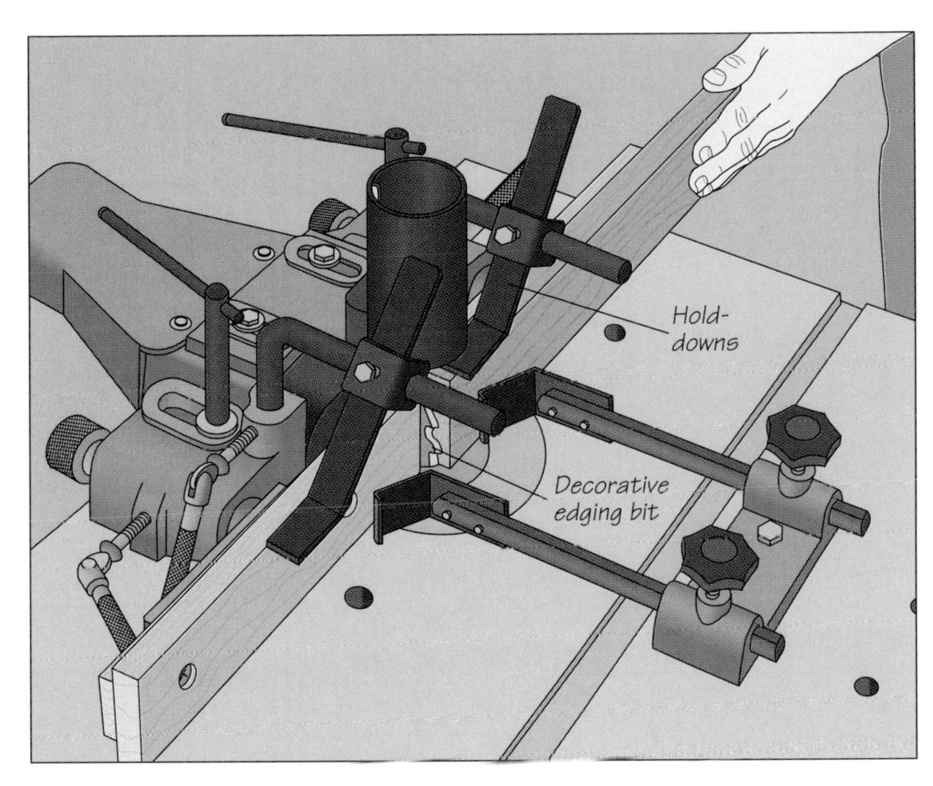

Before the router and shaper, the combination plane was the tool of choice for shaping molding. This versatile hand tool features a range of interchangeable cutters that can form tongues, grooves, dadoes, flutes, reeds, ovolos, and beadings. An adjustable edge guide ensures straight cuts while a depth stop allows the plane to trim to precise depths. The model shown at left, the Stanely 45 Multiplane, is a venerable design that inspired many imitators.

INSTALLING BASEBOARD

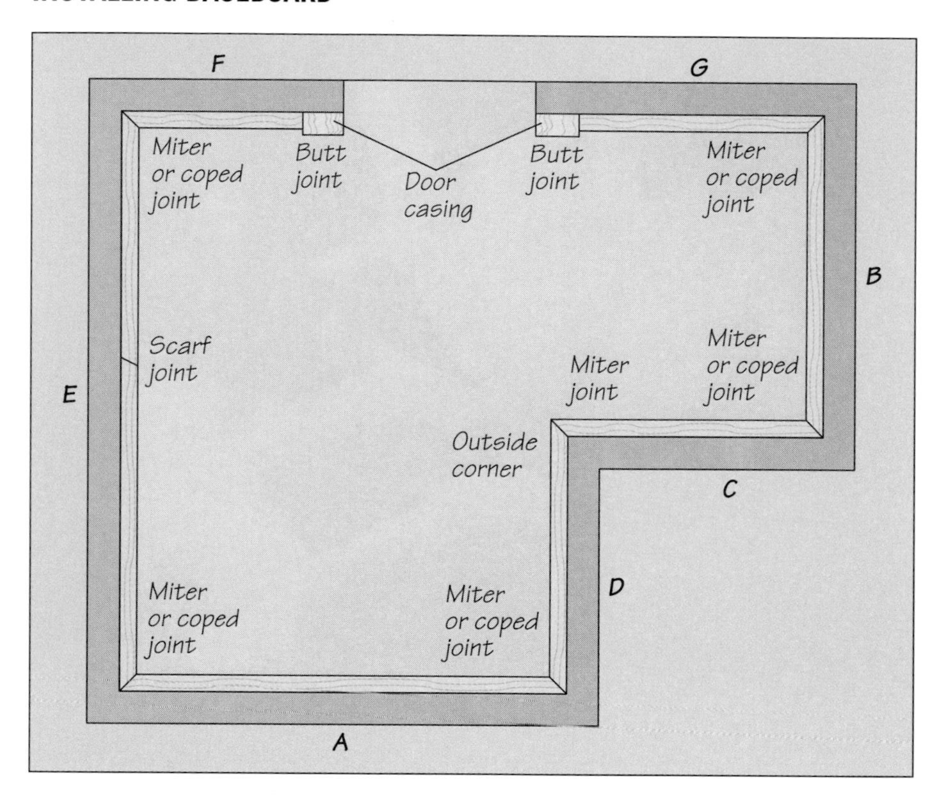

Planning the installation
The ideal sequence for installing baseboard depends on the room layout. Your goal is to make the joinery inconspicuous when entering the room, as shown at left. At door casings, use butt joints; at outside corners, use miters. At inside corners, use miters for flat molding *(below)* or coped joints for contoured molding *(page 27)*. Start at a long wall **(A)** opposite the door. With coped joints, cut the piece to butt against walls **D** and **E** so the end grain of the molding along these walls will be invisible from the door. Install the molding along wall **B** next, then walls **C** and **D**. If a wall is longer than your stock, as in **E**, connect two pieces using a scarf joint. Locate the joint at a wall stud. Finish the installation at the door (walls **F** and **G**). Cut all the molding 1/16-inch longer than needed; this will allow it to "snap" into place.

MITERING INSIDE CORNERS

Nailing the molding in place
To install base-and-shoe molding at an inside corner, cut the two pieces to length, mitering one end of each board *(page 17)*. Make the cuts so the back face of each molding reaches the corner, then install one of the pieces. Using a hammer and 2-inch (6d) finishing nails or a finish nailer loaded with the type of nails specified by the manufacturer, fasten the molding to the wall. Drive two nails at every wall stud, locating the nails ½ inch from the top and bottom of the molding. The upper nail should reach the stud, while the lower one should enter the sole plate attached to the subfloor directly below the studs. To locate the studs, use a stud finder *(page 32)*. If you are using a hammer, set the nail heads. Fit the second piece of molding in place *(right)* and nail it to the wall the same way. Then secure the shoe molding to the baseboard *(inset)*, driving a nail every 16 inches.

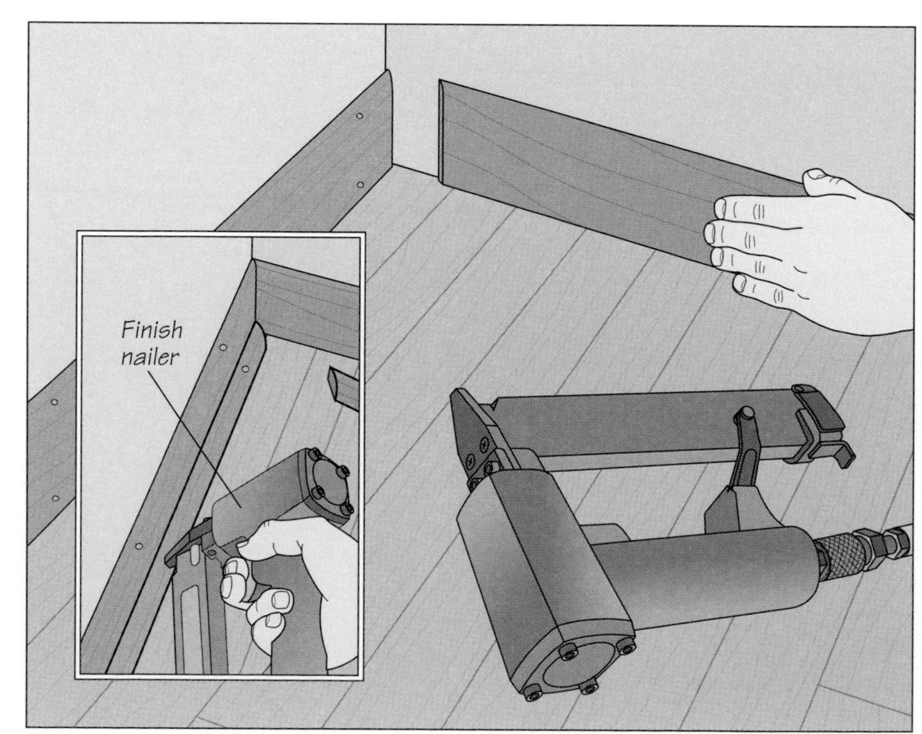

COPING INSIDE CORNERS

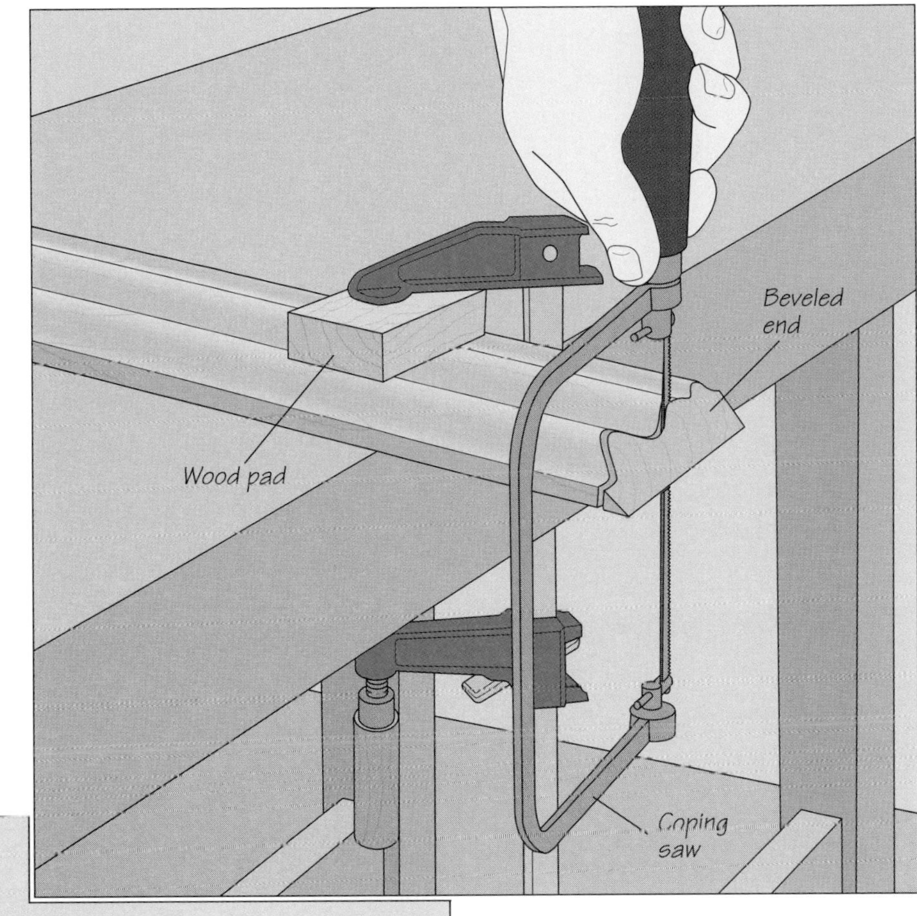

1 Coping the molding
To install contoured molding at an inside corner, crosscut both ends of one piece so that it fits snugly between the adjoining walls. The mating piece will butt against its face with a coped joint. Cutting this joint is a two-step operation. Start by making a 45° bevel cut on one end of the molding; this will reveal the contour line on the face. Then clamp the molding face up on a work surface, protecting the stock with a wood pad. Use a coping saw fitted with a narrow blade to cut along the contour line. Hold the saw perfectly upright *(right)*, biting into the wood on the upstroke. For a tight fit, hold the saw slightly over 90°, undercutting the joint slightly, so that only the front of the board contacts the face of the mating piece. If the blade binds in the kerf, make occasional release cuts into the waste to let small pieces fall away.

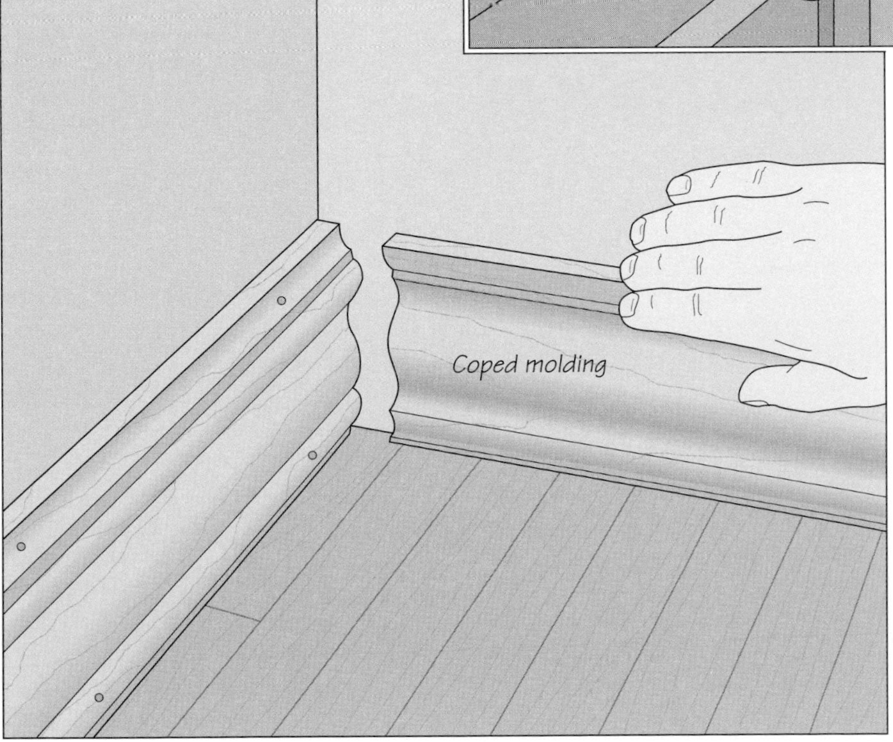

2 Installing the molding
Nail the first piece of molding to the wall as you would flat baseboard *(page 26)*. Then position the coped end against the first piece to test the fit *(left)*. Smooth out any irregularities with a round file or fine sandpaper wrapped around a dowel. Once the fit is perfect, nail the coped molding in place.

INSTALLING BASEBOARD AT AN OUTSIDE CORNER

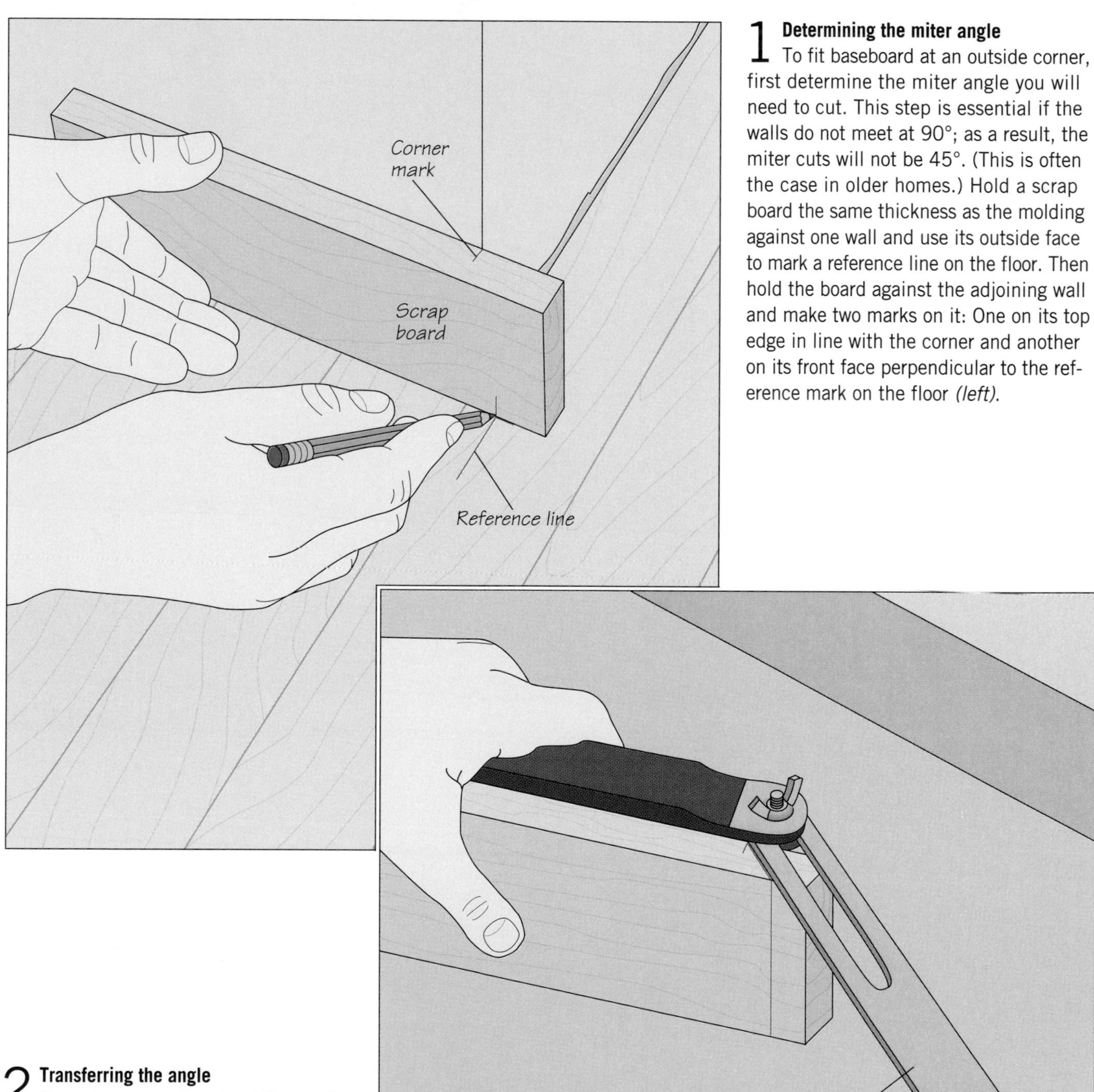

1 Determining the miter angle
To fit baseboard at an outside corner, first determine the miter angle you will need to cut. This step is essential if the walls do not meet at 90°; as a result, the miter cuts will not be 45°. (This is often the case in older homes.) Hold a scrap board the same thickness as the molding against one wall and use its outside face to mark a reference line on the floor. Then hold the board against the adjoining wall and make two marks on it: One on its top edge in line with the corner and another on its front face perpendicular to the reference mark on the floor *(left)*.

2 Transferring the angle
Use a try square to extend the mark on the face of the scrap board to the top edge. Then adjust a sliding bevel to the angle formed by the end of this line and the corner mark on the top edge of the board *(right)*. This is your miter angle. Use the sliding bevel to adjust the saw you will be using to cut the molding.

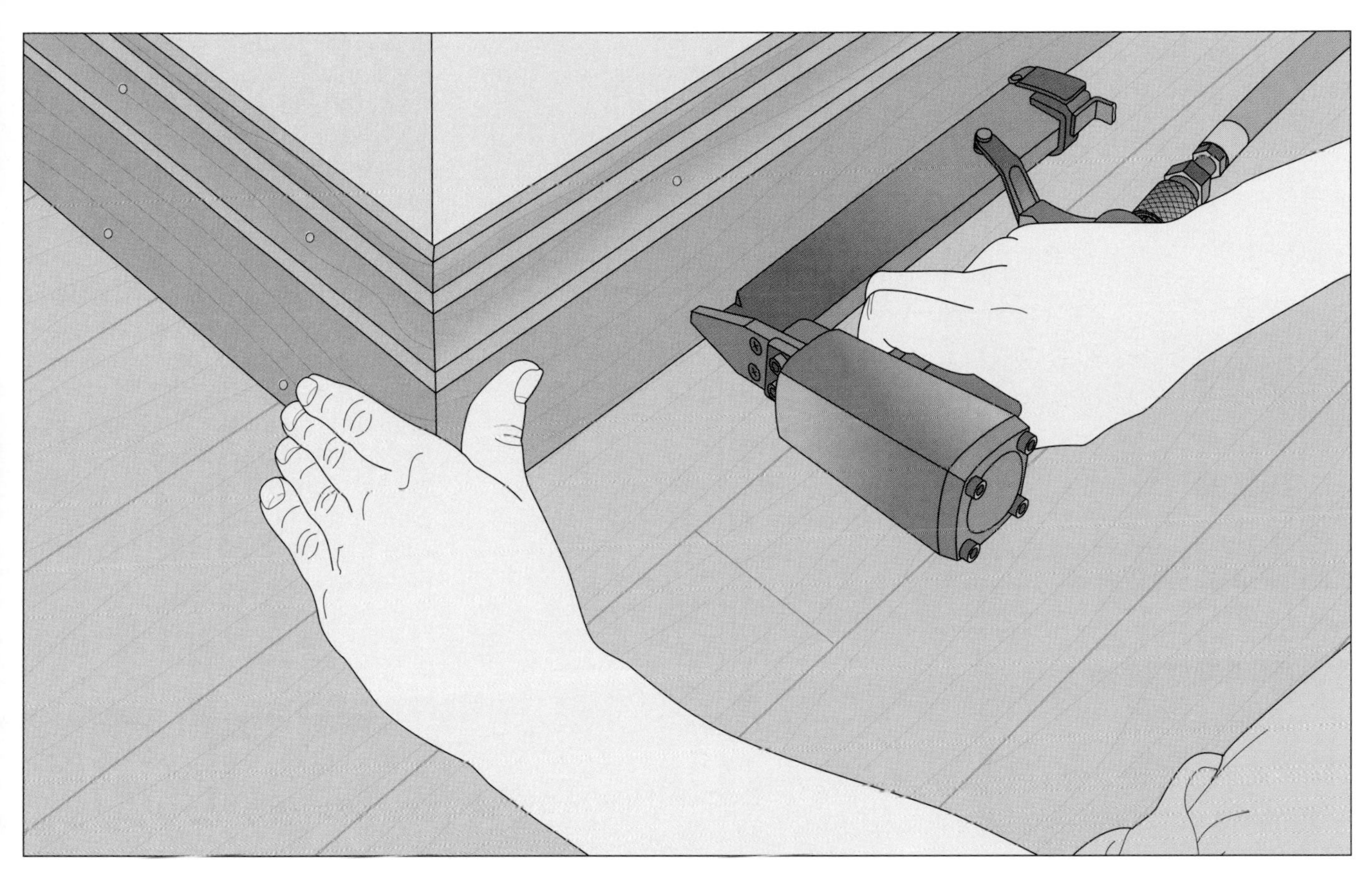

3 Installing the molding
Make the miter cuts on the pieces of molding and check the fit of the joint. Make any slight adjustments with a chisel or sandpaper. Then install the pieces with a hammer or finish nailer *(above)* as you would at an inside corner *(page 26)*. Repeat the process to install shoe molding or a base cap.

SHOP TIP

Store-bought corner pieces
Cutting a miter joint is not the only way to install baseboard at outside corners. Many types of commercial baseboard come with ready-made corner pieces featuring the same profile as straight sections. Joined to straight lengths with butt joints, these corners speed up installation. The square type shown here is made slightly "proud" of the straight sections for visual effect.

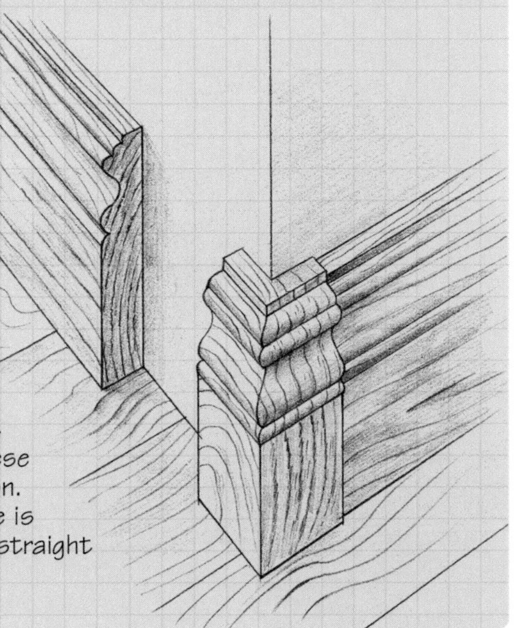

CHAIR AND PICTURE RAILS

Chair rails originally were used simply to prevent chair backs from marring walls or paneling, but they have recently assumed a more purely decorative role. Today, they often serve as a divider between decorative elements on a wall, with paneling or wallpaper usually installed between the chair rail and the floor.

Chair rails are much like baseboard, and are installed exactly the same way, except that they are located about one-third of the way between floor and ceiling, typically 3 feet off the floor. Like baseboard, chair rails are commercially available in a number of profiles and sizes, but they can be easily made in the shop from 1-by-4 stock.

Picture rail is a type of chair rail with a rounded lip used to hang picture frames. It is installed 6 to 8 feet off the floor. Since picture rails often support considerable weight, they are screwed rather than nailed in place. The screw holes are then concealed by wood plugs. Depending on the esthetic effect you want to achieve, you can use chair rails, picture rails, or both in conjunction with baseboard *(page 23)* and crown molding *(page 33)*.

Custom chair rails and crown molding are available with elaborate carved patterns and scrollwork, in a wide range of modern and antique styles.

MAKING CHAIR RAILS

1 Milling the reeds

Making one-piece molded chair rail typically involves two steps: cutting reeds on the face of the stock, as shown at right, and then shaping the edges *(step 2)*. Cut the reeds on your table saw using a molding head with a set of reeding knives. Install the head and position the rip fence by centering the workpiece face down over the cutters and butting the fence against the stock. Secure the workpiece with two featherboards, one clamped to the fence and a second fixed to the saw table. Both featherboards should be in line with the cutters. Clamp a support board at a 90° angle to the second featherboard. Make the first pass with a cutting height of ⅛ inch; do not make a full-depth cut in one pass. To make the cut, slowly feed the workpiece into the cutters with your right hand, pressing it against the rip fence with your left hand *(right)*. Finish the cut with a push stick. One or two passes is usually sufficient with this particular profile; raise the cutters no more than ⅛ inch at a time between passes.

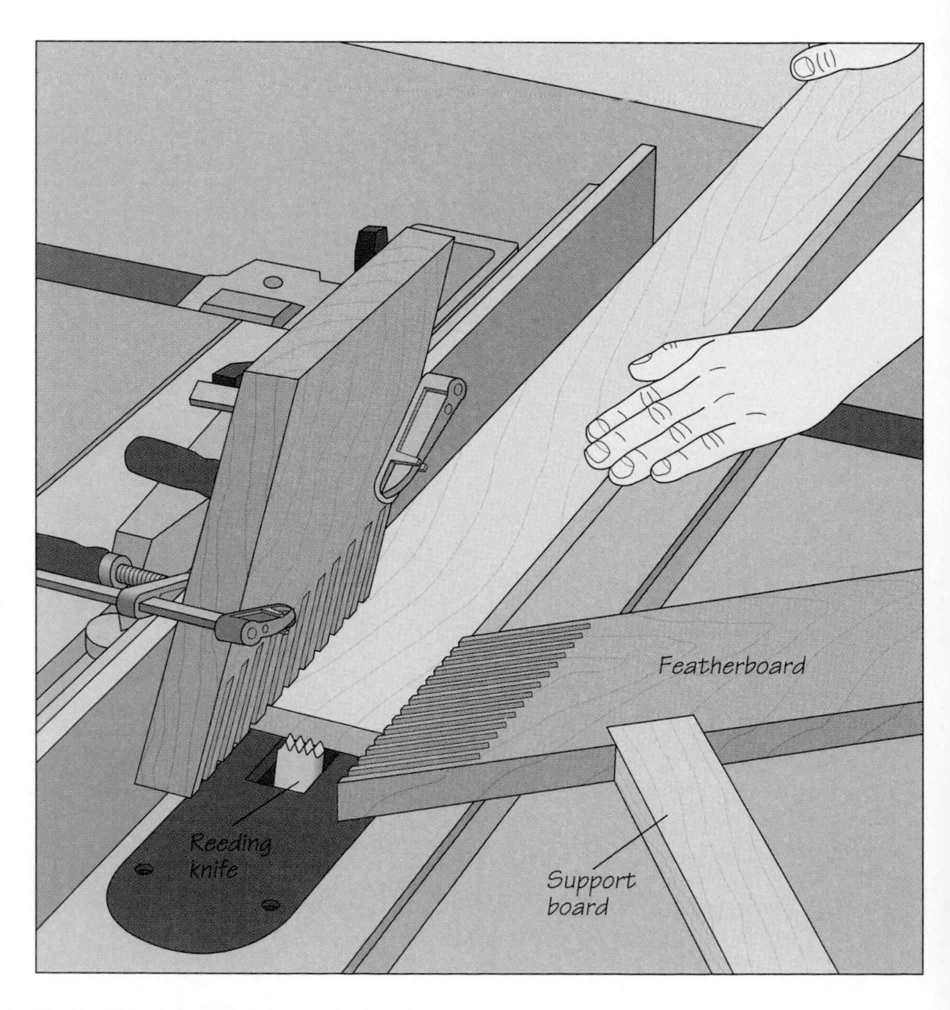

2 Shaping the edges
Once you have milled the reeds in your chair rail stock, shape the edges on a router table. Install an edge-forming bit in your router; an ogee bit is shown at right. Then mount the tool in a table. To support the workpiece for this cut, use three featherboards: Clamp two featherboards to the fence, one on each side of the bit, and use the third opposite the bit to press the stock against the fence. Set a shallow depth of cut for the first pass. To shape each edge, slowly feed the workpiece into the cutter *(right)*; finish the cut with a push stick. Make a series of deeper passes until you attain the desired profile, increasing the cutting depth ⅛ inch at a time.

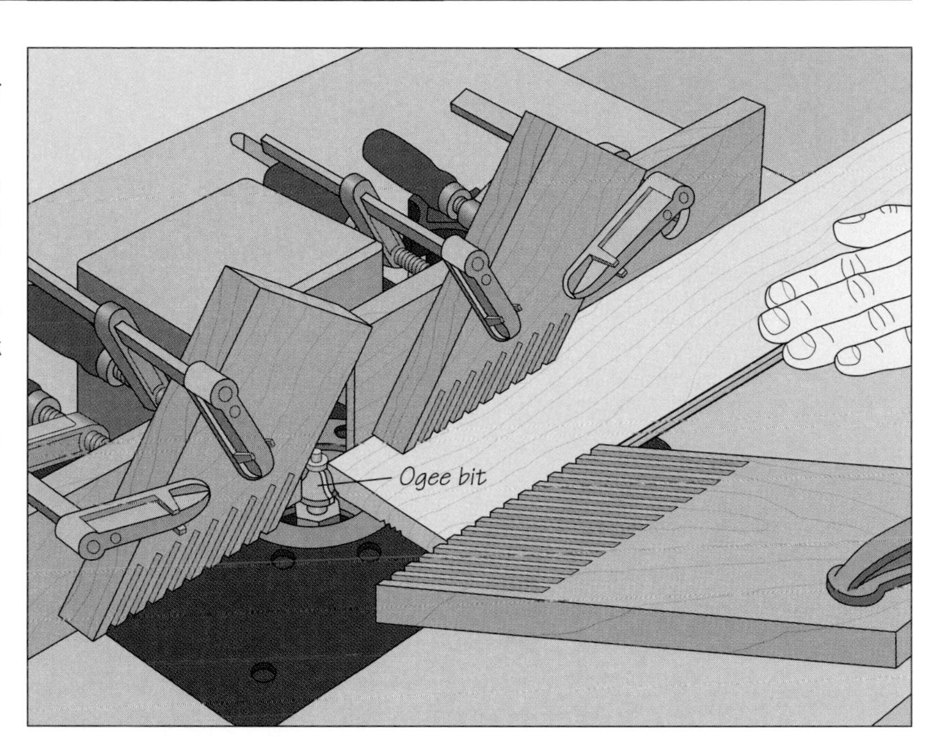

INSTALLING A CHAIR RAIL

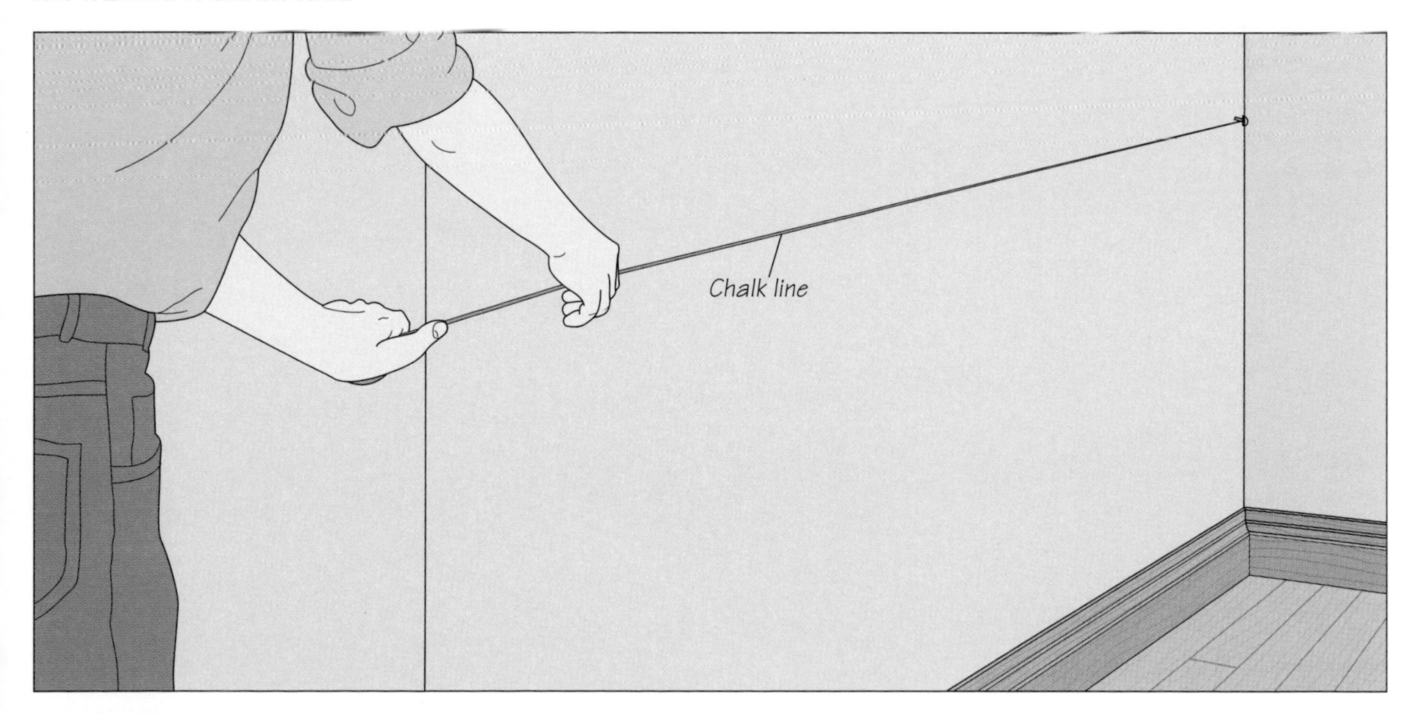

1 Determining the height of the chair rail
Use a chalk line to mark a height line on the wall for the molding. Make a mark at each end of one wall, typically 36 inches off the floor. After measuring the height of one mark, use a level and a long, straight board to make the second mark at the same level, as the floor may not be true. Drive a finishing nail into the wall at one of the marks and hook the chalk line on the nail head. Align the other end with the second height mark and snap the chalk line *(above)*. Repeat for the other walls in the room.

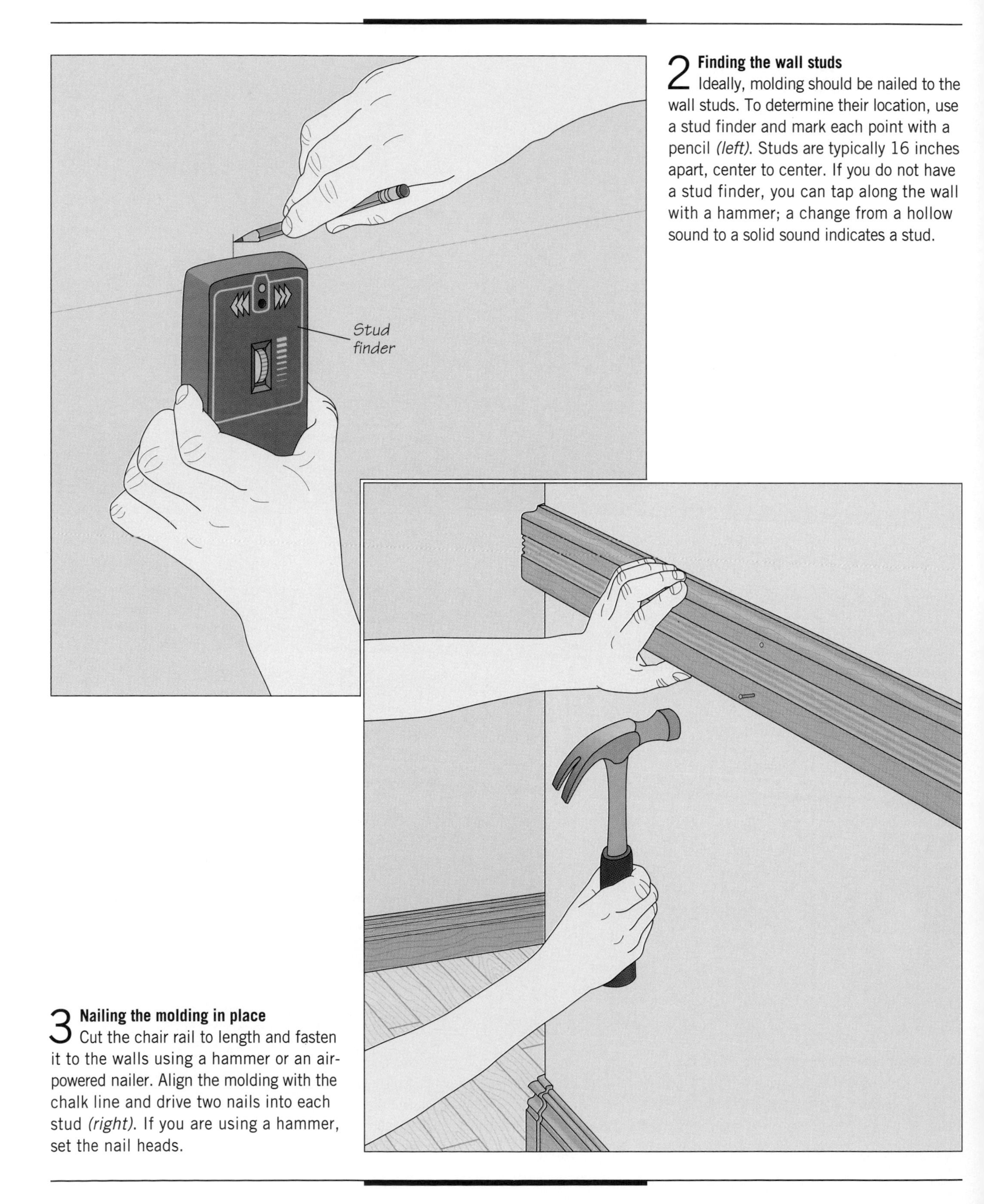

2 Finding the wall studs
Ideally, molding should be nailed to the wall studs. To determine their location, use a stud finder and mark each point with a pencil *(left)*. Studs are typically 16 inches apart, center to center. If you do not have a stud finder, you can tap along the wall with a hammer; a change from a hollow sound to a solid sound indicates a stud.

3 Nailing the molding in place
Cut the chair rail to length and fasten it to the walls using a hammer or an air-powered nailer. Align the molding with the chalk line and drive two nails into each stud *(right)*. If you are using a hammer, set the nail heads.

CROWN MOLDING

Rooted in classical Greek architecture, crown molding rose to prominence during 18th-Century England, first appearing on furniture of the Georgian period. Soon after, the flowing patterns of this molding also began to adorn the ceilings of drawing rooms. Today, single-piece crown molding can be installed along the walls and ceilings of house interiors, or it can be combined with other elements to make built-up ceiling moldings, such as crown-and-flat *(page 22)*, or a formal cornice *(page 36)*. When choosing crown molding, make sure it is properly proportioned for the room; molding that is too wide will give the effect of lowering the ceiling. Molding 3 or 4 inches in width is about right for an average-sized 8-foot-high ceiling.

Installing crown molding is not much different from nailing on baseboard or chair rails; outside corners are mitered and inside corners are mitered or coped. Simple one-piece crown molding is nailed through the flats of the molding into the wall studs, ceiling joists, and top plate, which rests on top of the wall studs. Complex built-up crown molding, such as a formal cornice, needs to be fastened to furring strips. Where the ceiling joists run parallel to the wall, gluing the molding to the ceiling will often suffice. A pair of nails driven at opposite angles into the wall will hold the molding in place until the adhesive cures.

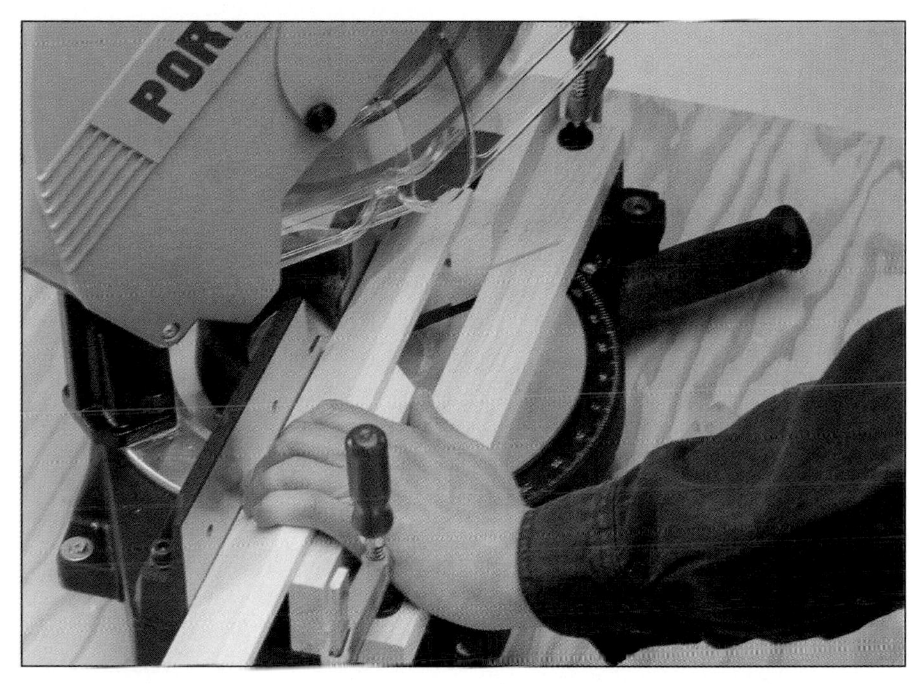

Because crown molding is positioned on both the wall and ceiling, corners must be mitered at compound angles, and pieces must be held "upside down and backward" when they are cut. The jig shown in the photo above, however, allows the molding to be held on the chop saw exactly as it will appear on the wall and ceiling, requiring no repositioning before the cut.

INSTALLING ONE-PIECE CROWN MOLDING

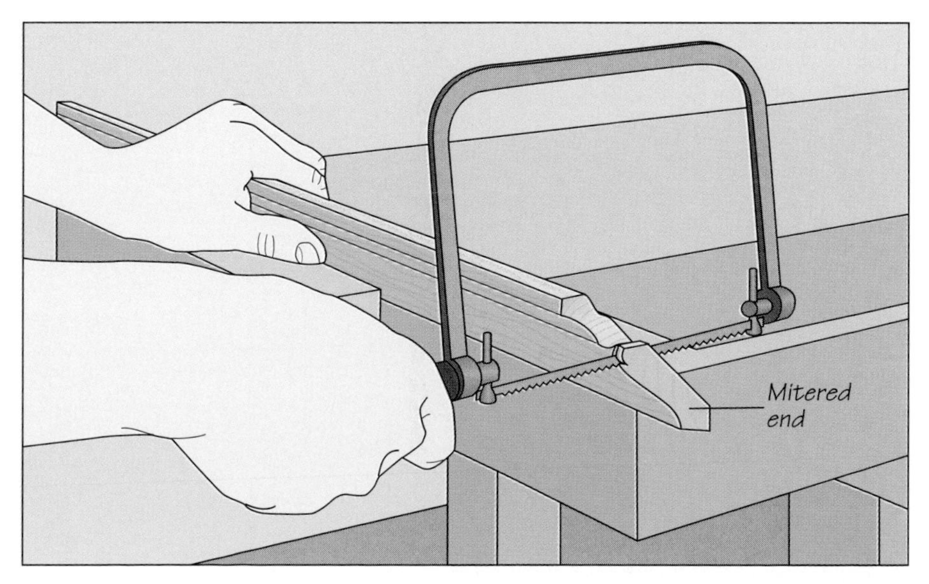

1 Coping molding at an inside corner
Before installing crown molding, place it in position, and snap a chalk line on the wall to mark the bottom edge of the molding. Cope the end of one piece to fit against the face of an adjoining piece as you would baseboard *(page 27)*. Start by mitering the end, then make the coped cut using a coping saw. In this case, secure the workpiece in a vise in the same position it will be when it is installed. Hold the coping saw perpendicular to the molding throughout the cut *(left)*.

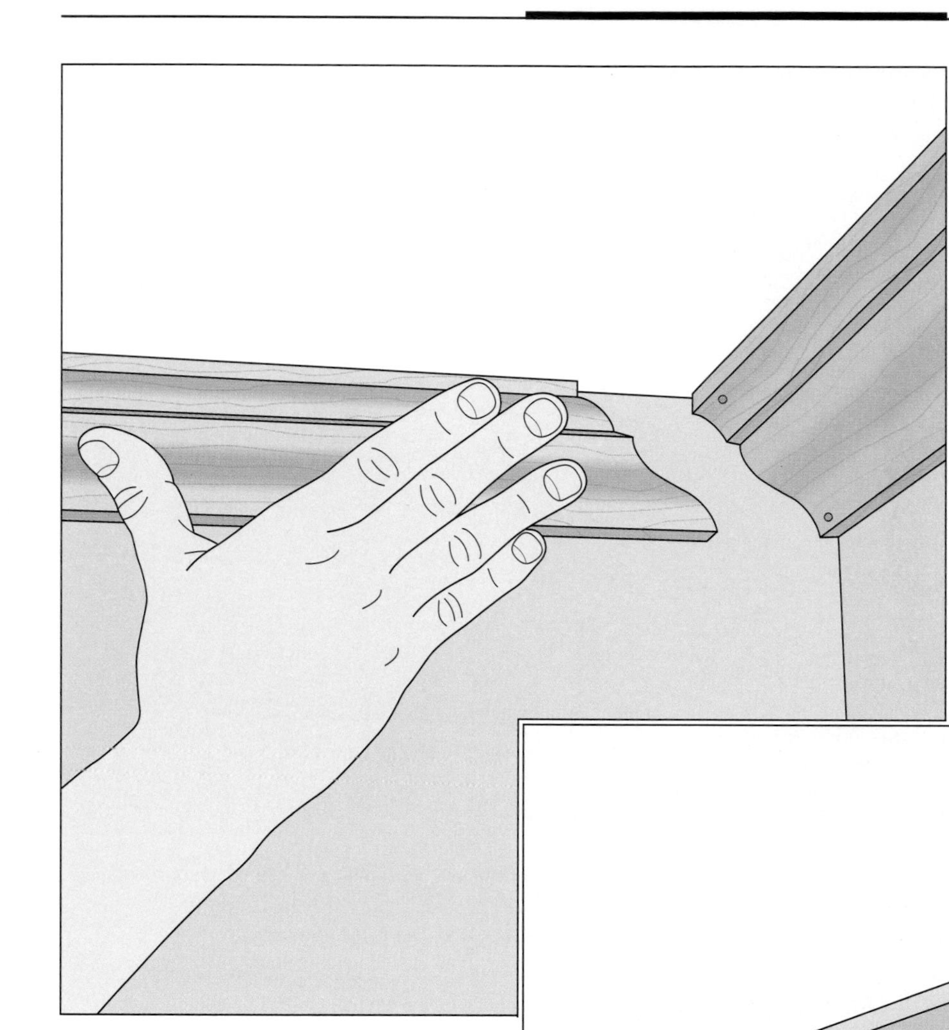

2 Installing the coped molding
Test-fit the coped end against the adjoining piece of molding, which should already be nailed in place *(left)*. Fine-tune the fit, if necessary, by filing or sanding the coped end. Nail the coped molding to the wall and ceiling as you would base-and-shoe molding *(page 27)*.

Reference lines

3 Determining the miter angle at an outside corner
As with baseboard, the first step in installing crown molding at an outside corner is to find the correct miter angle. Start by drawing two reference lines on the ceiling. Holding a piece of molding in place against one wall and the ceiling, mark one of the lines along its top edge, extending past the corner. Repeat for the adjacent wall to mark the second line *(right)*.

4 Transferring the miter angle to a sliding bevel

Draw a line from the corner to the point where the two reference lines you marked in step 3 intersect. Then adjust a sliding bevel so that its handle butts against one of the walls and the blade aligns with the line you just marked on the ceiling *(right)*. Use the sliding bevel to set up your saw to cut the miters.

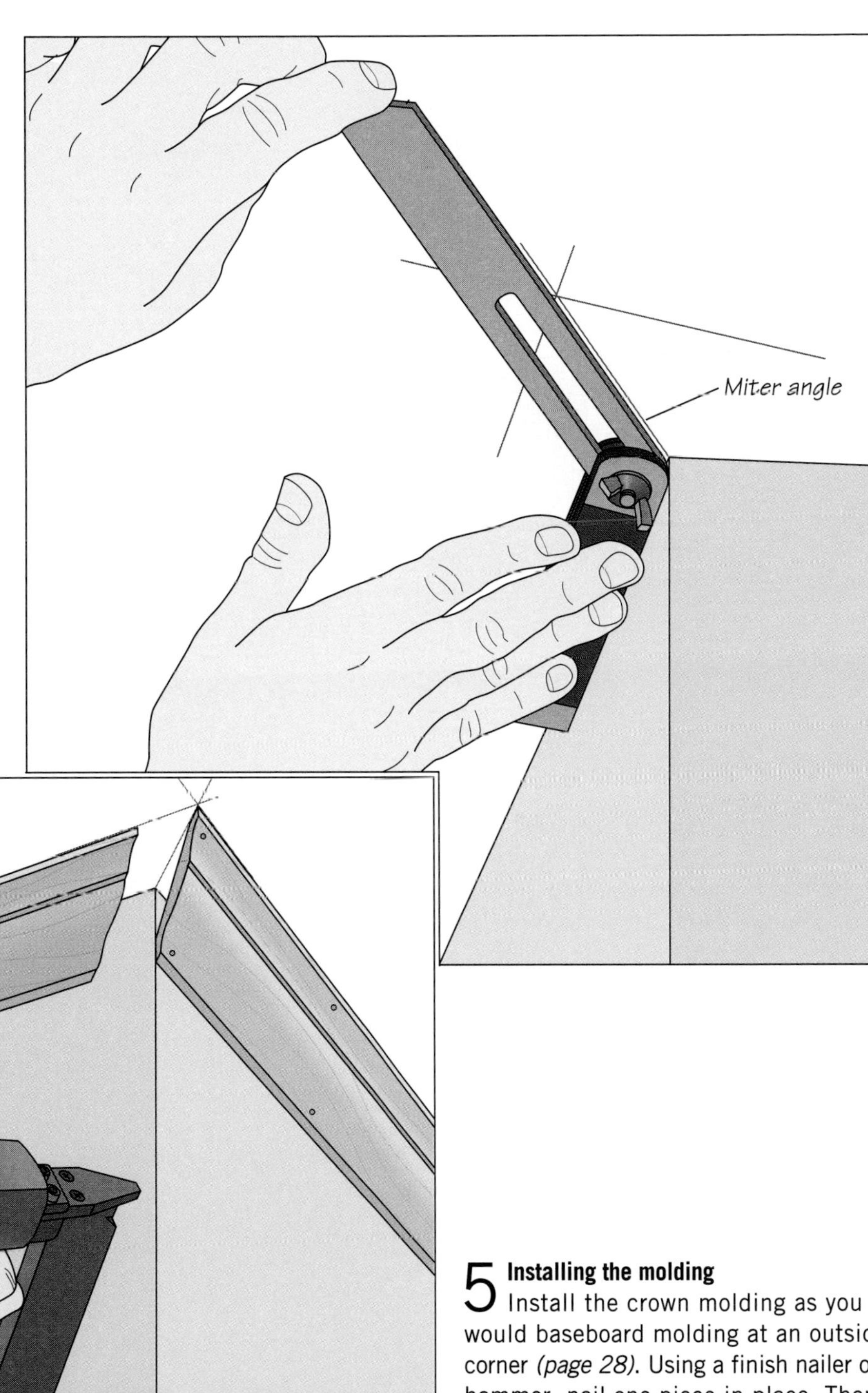

5 Installing the molding

Install the crown molding as you would baseboard molding at an outside corner *(page 28)*. Using a finish nailer or hammer, nail one piece in place. Then position the other *(left)*. If the fit is not perfect, back cut the miters slightly with a utility knife. Once you are satisfied with the fit, apply some glue to the mitered ends and nail the second piece of molding in place. To prevent the miter from opening, nail through the miter from both sides.

INSTALLING A FORMAL CORNICE

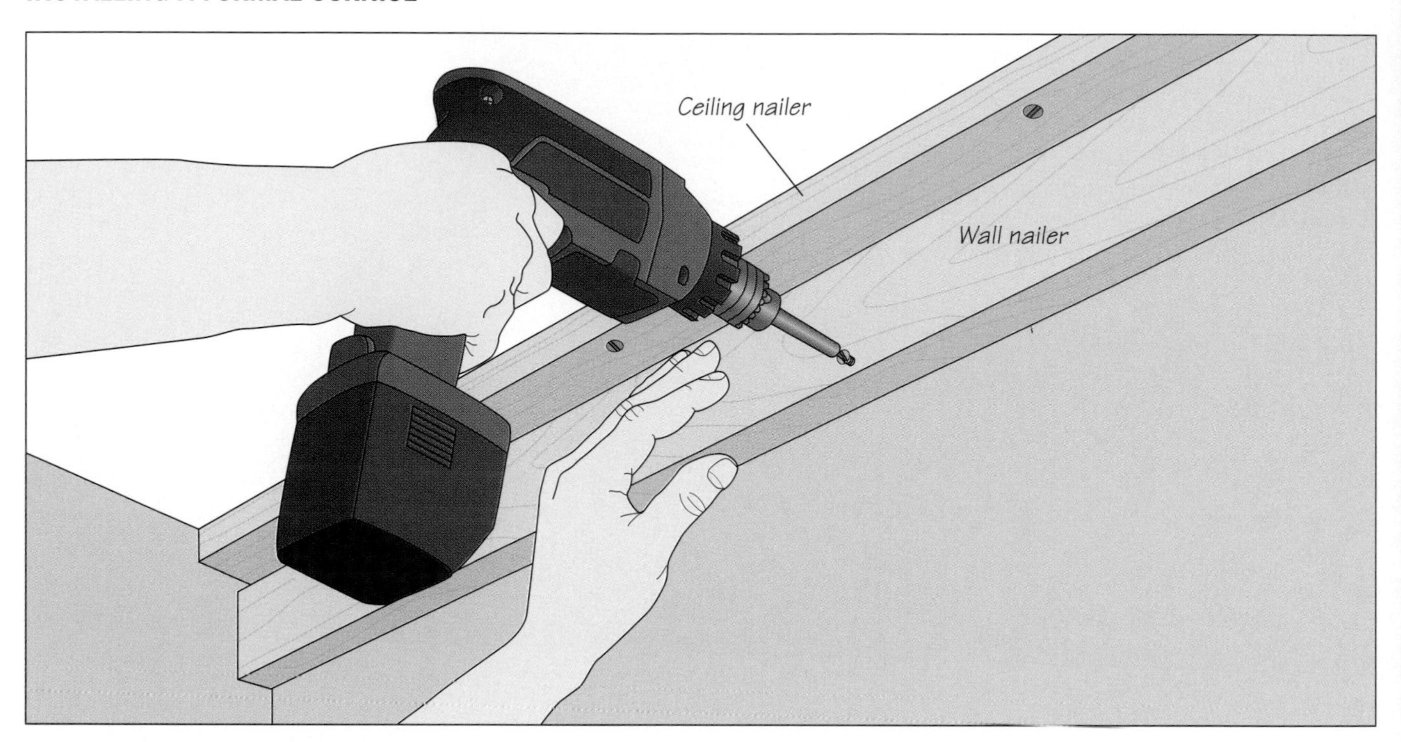

1 Installing nailing strips
A formal cornice is an antique-style crown molding consisting of a box-like support assembly (installed in steps 1 and 2) and three pieces of molding (installed in step 3). Cut the four pieces of the support assembly from ¾-inch stock. These pieces are installed the same way as baseboard *(page 26)*, with miters at both inside and outside corners. Start by screwing the ceiling nailer into the ceiling joists with the piece flush against the wall. Then screw the wall nailer to the wall studs *(above)*, leaving a ⅛-inch gap between its top edge and the ceiling nailer to allow for wood movement.

2 Installing the soffit and fascia
Rip the fascia and soffit pieces so that when they are attached to the ceiling and wall nailers, the four pieces will form a box. Rout a decorative lip along the bottom edge of the fascia piece, join the fascia and soffit boards with plate joints, then screw the fascia to the ceiling nailer and attach the soffit to the wall nailer *(right)*.

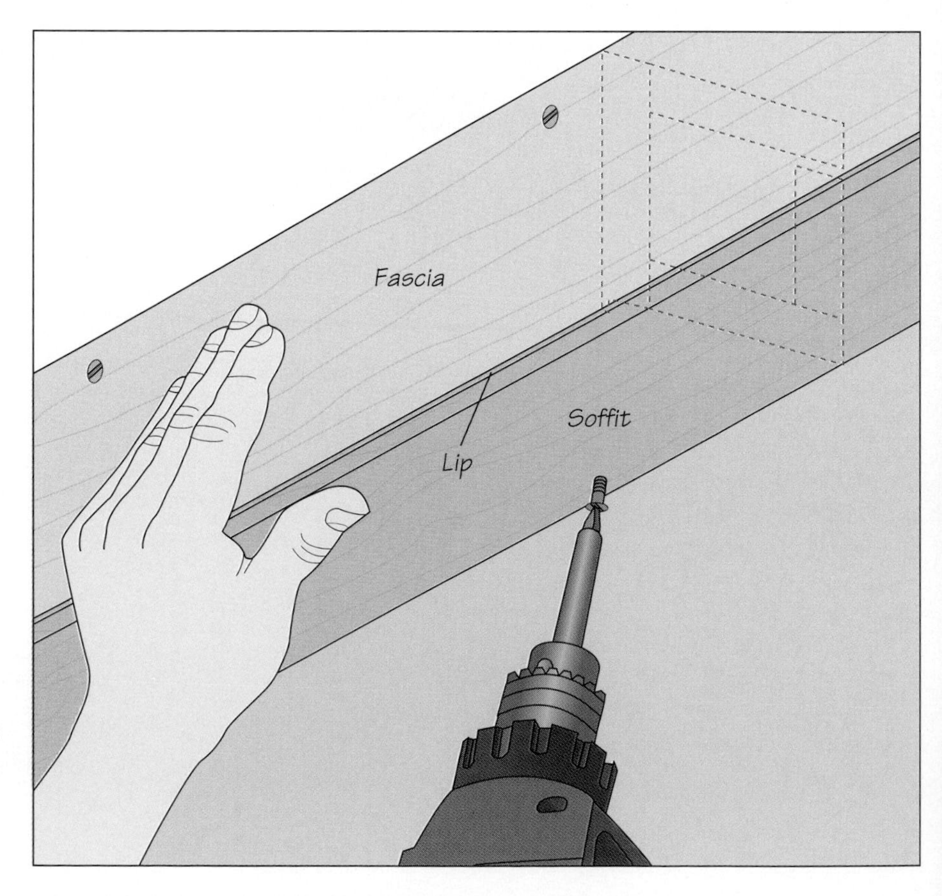

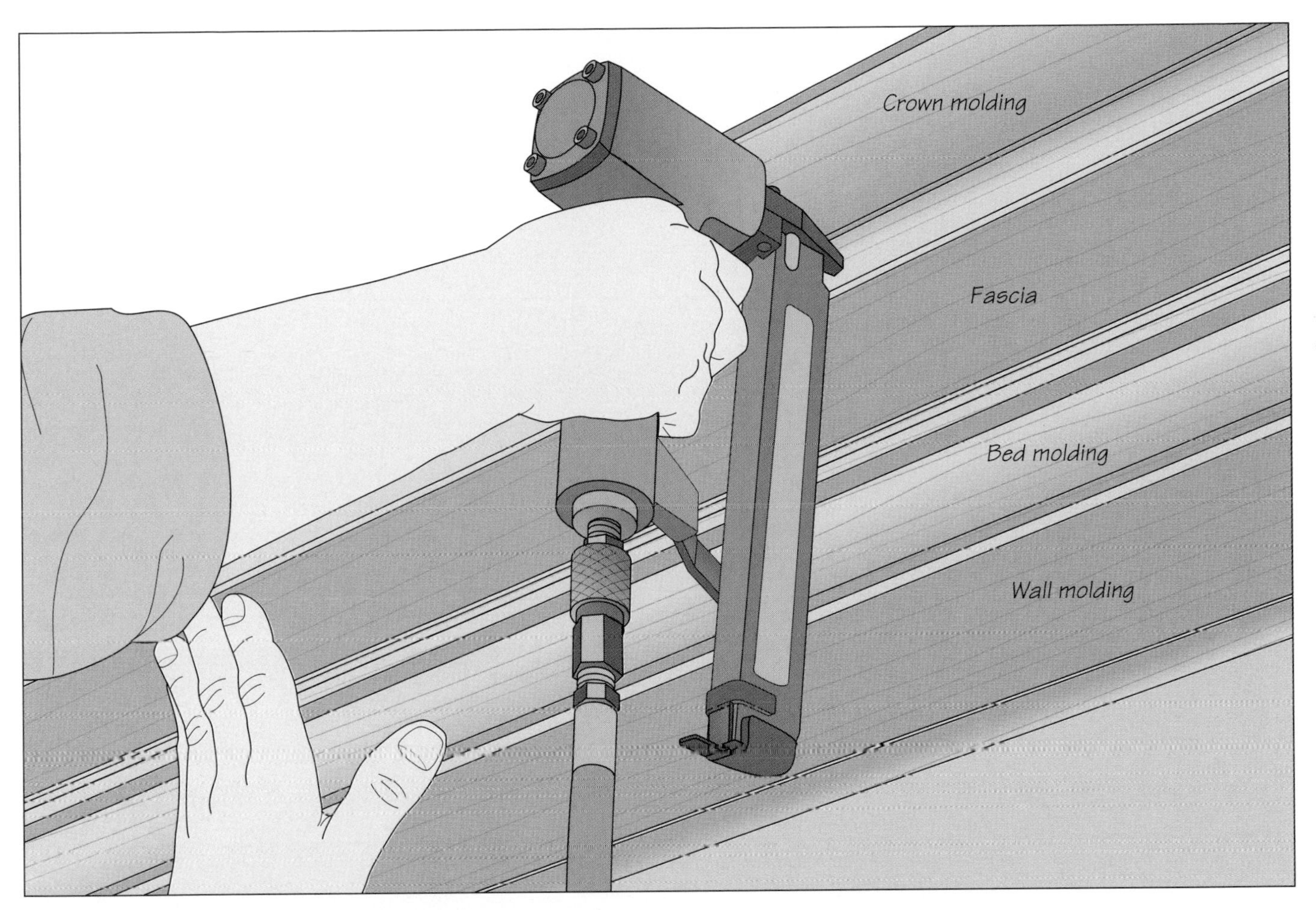

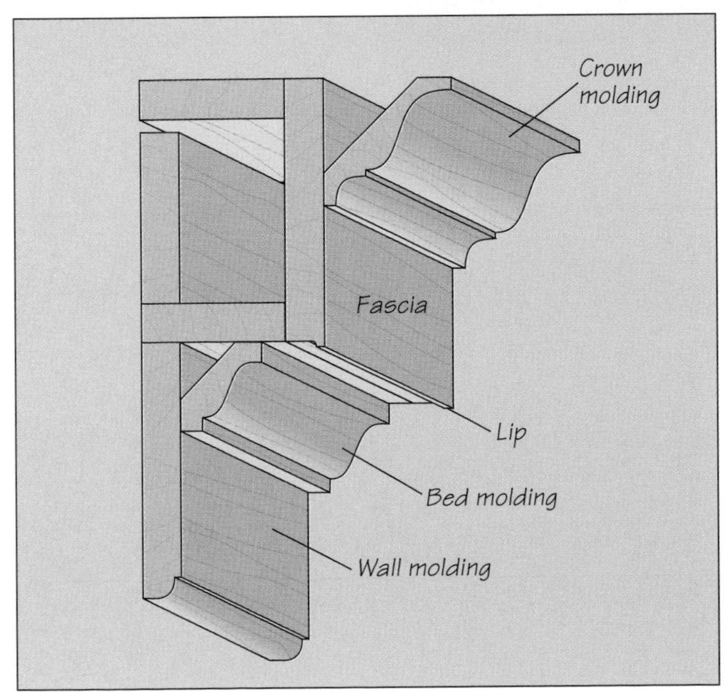

3 Installing the molding

To make the wall molding, rout a decorative edge in ¾-inch stock, then screw the piece in place with its flat edge against the soffit. Drive the screws into the wall studs near the molding's top edge, where the fasteners will be concealed by the bed molding; nail along the bottom edge. Next, install the bed molding as you would crown molding *(page 33)*, nailing it to the wall molding and the soffit. Finally, cut and install the crown molding between the fascia and the ceiling around the room *(pages 33-35)*. Nail the molding through its flat sections, driving the fasteners into the ceiling joists and the fascia *(above)*.

PANELING

Interior walls have been graced by frame-and-panel wainscoting for centuries. The flexibility of frame-and-panel construction allows for a wide range of design possibilities, from the elaborate ornamental panels of Jacobean-style furniture to more modern designs, such as the full-wall paneling shown in the photo above, made by Patella Industries of Montreal.

Since the late Middle Ages, wood paneling has been used to beautify interior walls. Often designed and colored to complement the furniture in a room, paneling warms a room's ambiance, giving it depth and character. The most popular and enduring form of interior paneling is wainscoting. In 16th-Century Europe, at the end of the Gothic furniture period, "waynscottes" consisted of seasoned planks of Baltic oak covering interior walls. In Colonial America, "wainscot" referred to pine boards that stretched from floor to ceiling, adding a rustic warmth to parlors. Today, wainscoting encompasses a wide range of wall coverings, although the term most frequently describes panels installed on the lower half of a wall.

There are two basic types of paneled wainscoting: tongue-and-groove and frame-and-panel. Tongue-and-groove wainscoting *(page 42)* is available at hardware stores and lumberyards as ready-to-install interlocking paneling. Made from stain-grade hardwood or paint-grade softwood, tongue-and-groove paneling comes in a variety of profiles. But you can easily make your own with a table saw, some 1-inch-thick stock, and a molding cutterhead, creating styles that often are not available commercially.

Frame-and-panel wainscoting *(page 46)* is a modular system in which beveled panels lie in a frame of vertical members, called stiles or mullions, and horizontal ones, called rails. Grooves are milled around the inside edges of the frame, allowing the panels to float freely, expanding and contracting with seasonal changes in humidity. This custom wainscoting can transform a plain room into a more formal space, replacing flat, monotonous walls with exquisitely molded panels.

Paneling techniques can also be adapted to ceilings to create a sumptuous look for a den or study *(page 52)*. In a paneled—or coffered—ceiling, a framework of 2-by-4s is sheathed in hardwood, with veneered plywood panels set into the frames. Crown and shoe molding provide the final decorative touch.

In any style, paneled surfaces can transform ordinary rooms into richly comforting retreats.

A molded cap rail is installed as the crowning touch to frame-and-panel wainscoting. In addition to its decorative role, the cap rail hides the gap between the wainscoting and the wall. Cap rail designs range from the simple chamfered rail shown at left to more elaborate molded chair rails.

PANELING STYLES

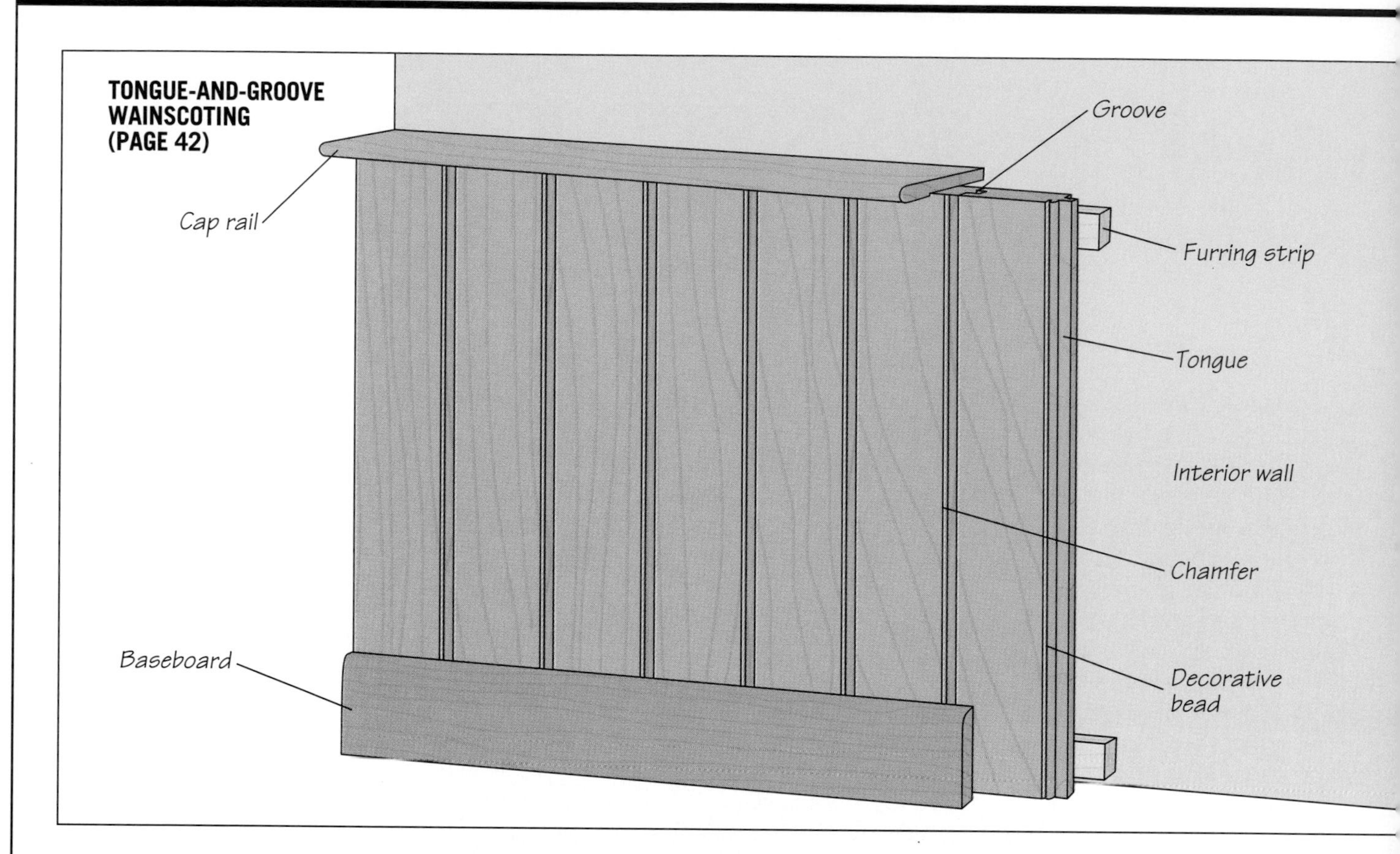

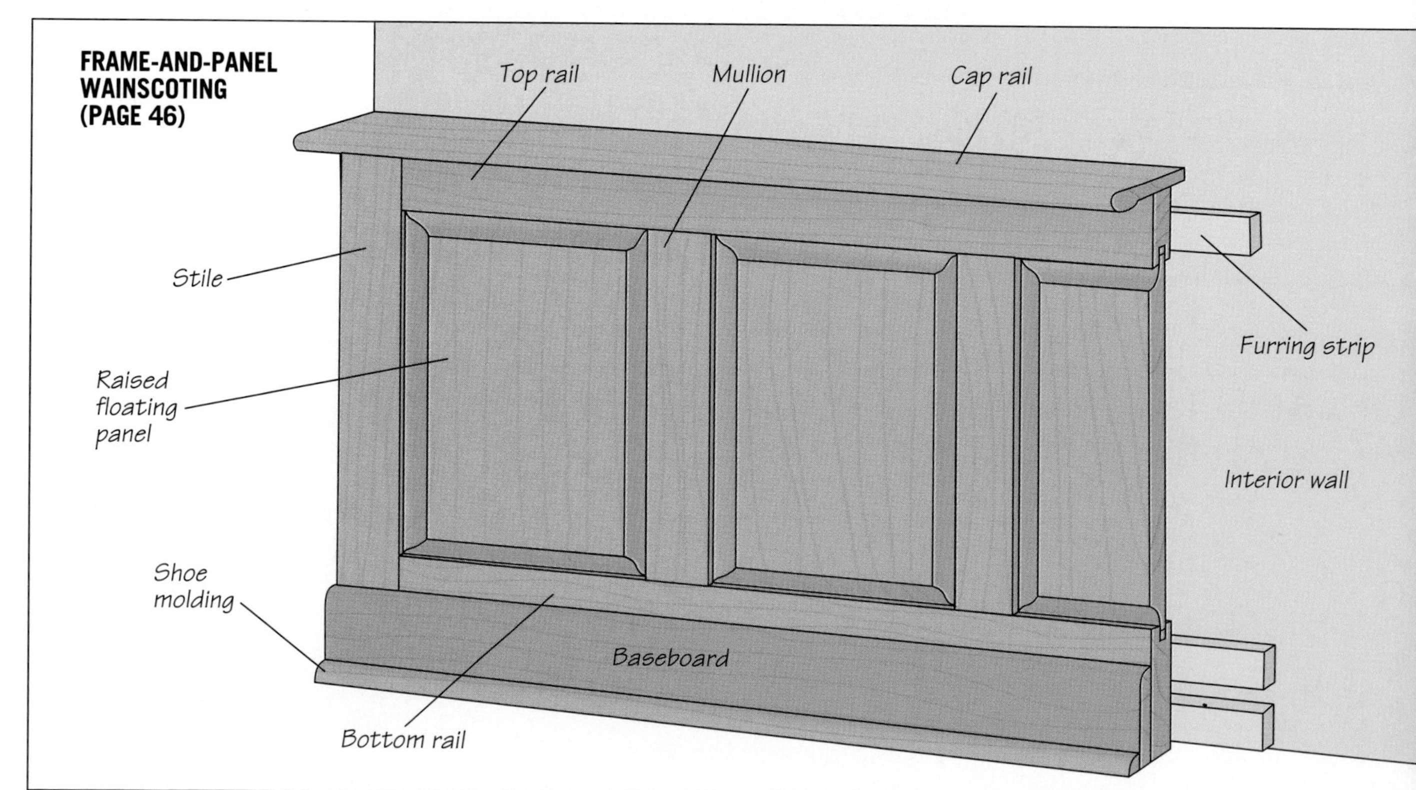

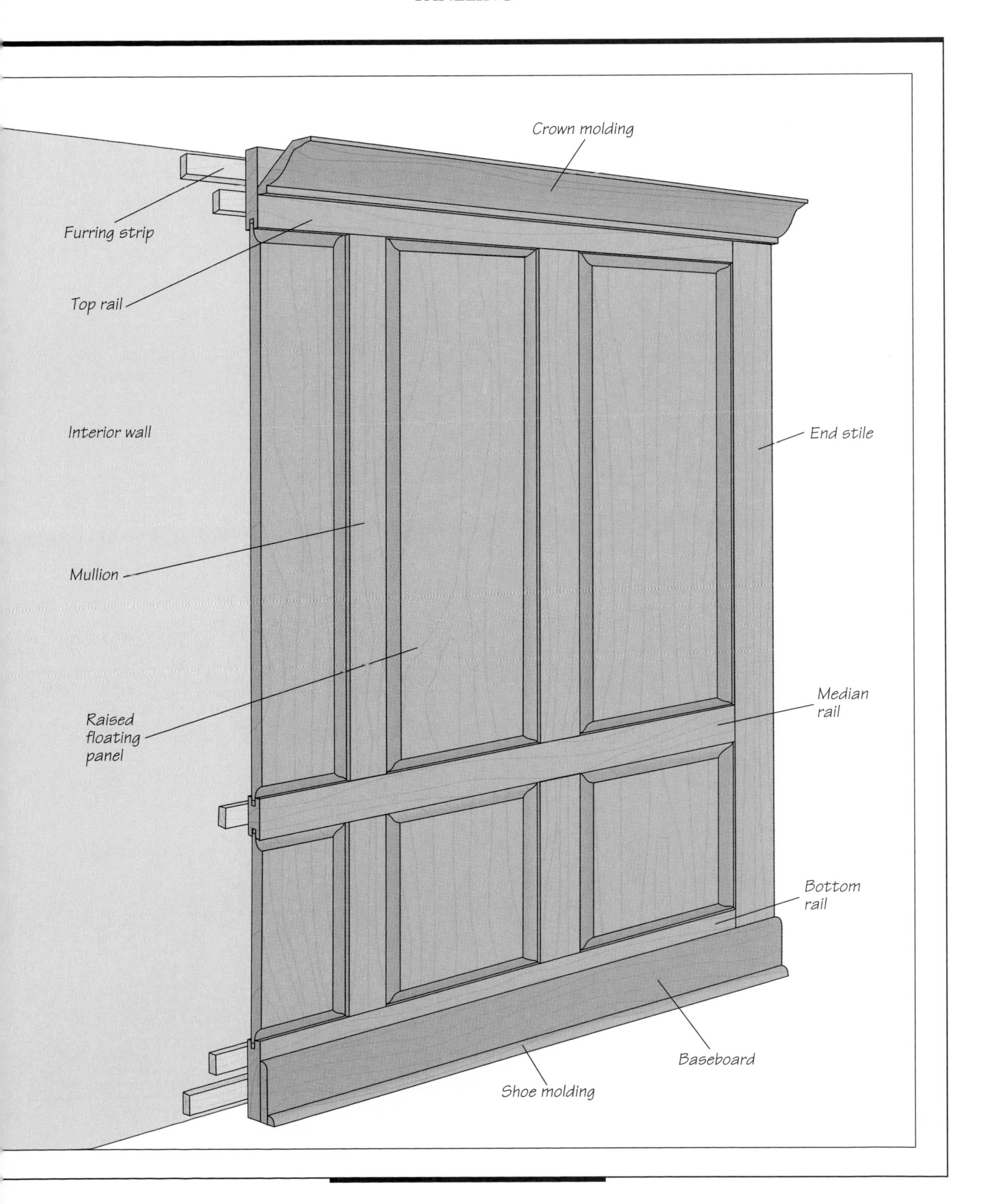
Crown molding
Furring strip
Top rail
Interior wall
End stile
Mullion
Raised floating panel
Median rail
Bottom rail
Baseboard
Shoe molding

TONGUE-AND-GROOVE WAINSCOTING

Tongue-and-groove wainscoting consists of a series of interlocking boards or panels with an optional decorative profile. It is the most basic and popular type of interior paneling. Traditionally made from softwood and given a clear finish, tongue-and-groove paneling can be installed from floor to ceiling to create a rustic look. It imparts a more sophisticated appearance when used as wainscoting and stained.

While you can buy tongue-and-groove wainscoting ready-made, you can also mill your own from 1-by-4 or 1-by-6 stock using a table saw. Use a dado head to cut tongues in one edge of the boards and grooves in the other edge. Switch to a molding head to mill a decorative bead in the front face of the boards; a selection of wainscoting profiles is displayed at right.

Tongue-and-groove wainscoting is typically installed from the floor to a height of 36 inches. Compensate for uneven floors by cutting the boards a little short. That way, the top ends of the boards can all be installed at the same level; any gaps between the bottom ends and the floor will be concealed by baseboard *(page 22)*. If there are nailers behind the wall—typically 2-by-4 blocking between the wall studs—and you know where they are, fasten the wainscoting to them. Otherwise, anchor furring strips to the studs *(page 49)* and nail the wainscoting to the furring strips.

A SELECTION OF PANEL STYLES

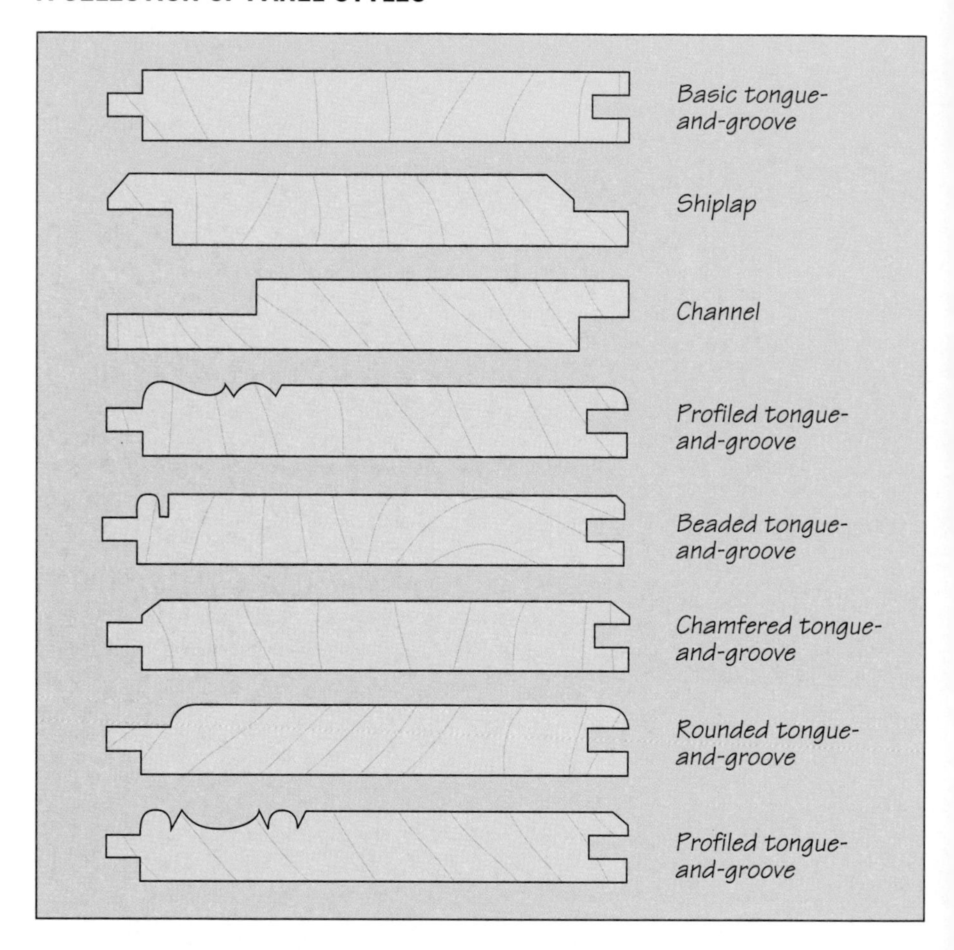

Butting the edge of a board or panel squarely against an out-of-plumb wall, brickwork, or a wall with contoured molding can be a challenge. The log-builder's scribe shown in the photo at right makes cutting the edge easy. It features an adjustable curved steel pin and two level vials for accurately tracing the wall profile onto the paneling.

MAKING TONGUE-AND-GROOVE WAINSCOTING

1 Cutting the grooves
Outline the groove on the leading end of one board; make the groove width one-third the stock thickness and its depth about ½ inch. Install a dado head on your table saw, adjusting its width to that of the groove and the cutting height to the groove depth. Install an auxiliary wood fence and cut a relief notch in it to house part of the dado head when you mill the tongues in step 2. Align the cutting marks with the dado head and butt the fence against the stock. To prevent the workpiece from tipping during the cut, clamp a shimmed featherboard to the saw table in line with the dado head. Secure a support board against the featherboard for extra pressure. Press the workpiece against the fence and the table as you feed it edge down into the dado head *(right)*. Complete the pass with a push stick. Use the same setup to cut the grooves in all the boards.

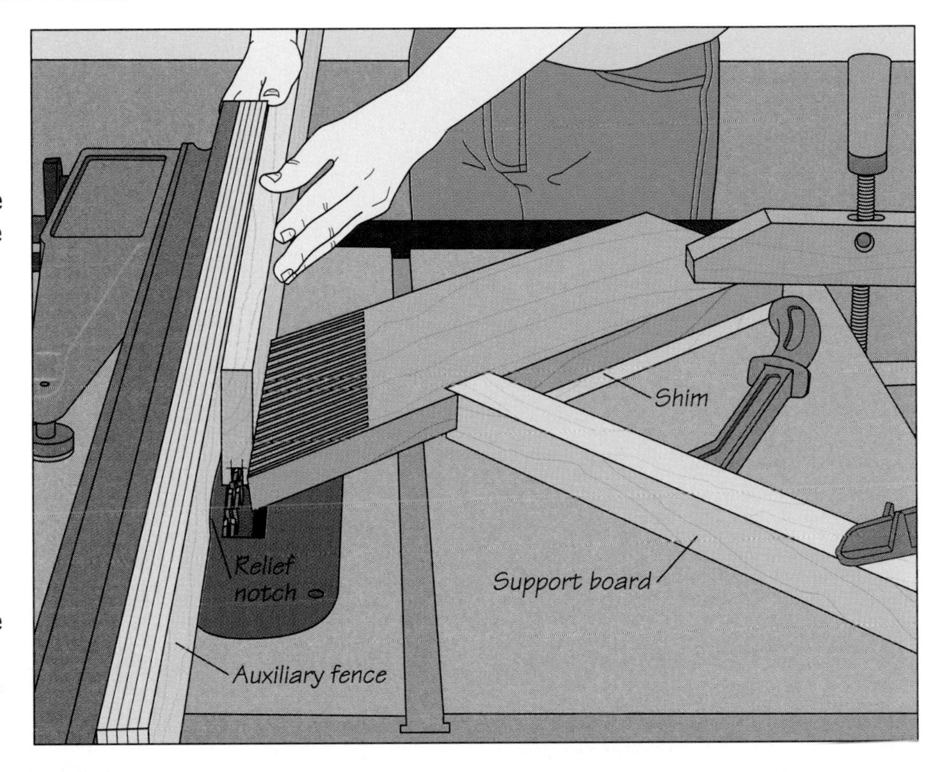

2 Cutting the tongues
Outline the tongue on the leading end of one board, using a groove you cut in step 1 as a guide. Lower the dado head slightly so the tongue will not bottom out in the groove. Then align one of the cutting marks with the cutters and butt the fence against the stock; also reposition the featherboard. Feed the board as you did for cutting the groove, using a push stick to finish the pass. Turn the workpiece around and repeat the cut to complete the tongue *(above)*.

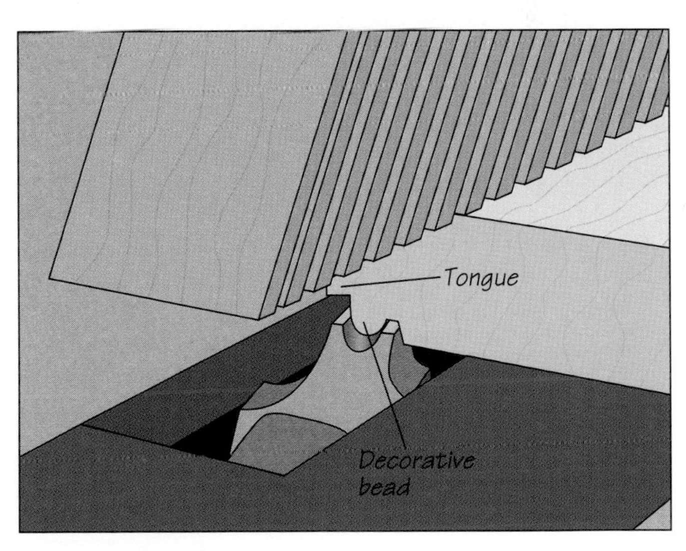

3 Milling the beads
Install a molding head fitted with beading knives on your table saw. Align the board face down over one of the knives so the bead will be milled alongside the tongue. Butt the fence against the board, reposition the featherboard, and clamp a second featherboard to the fence directly over the knives. Make a series of test cuts in a scrap board to determine the proper depth, then mill the beads *(above)*. You can also cut a slight chamfer in the opposite edge with a hand plane.

INSTALLING TONGUE-AND-GROOVE WAINSCOTING

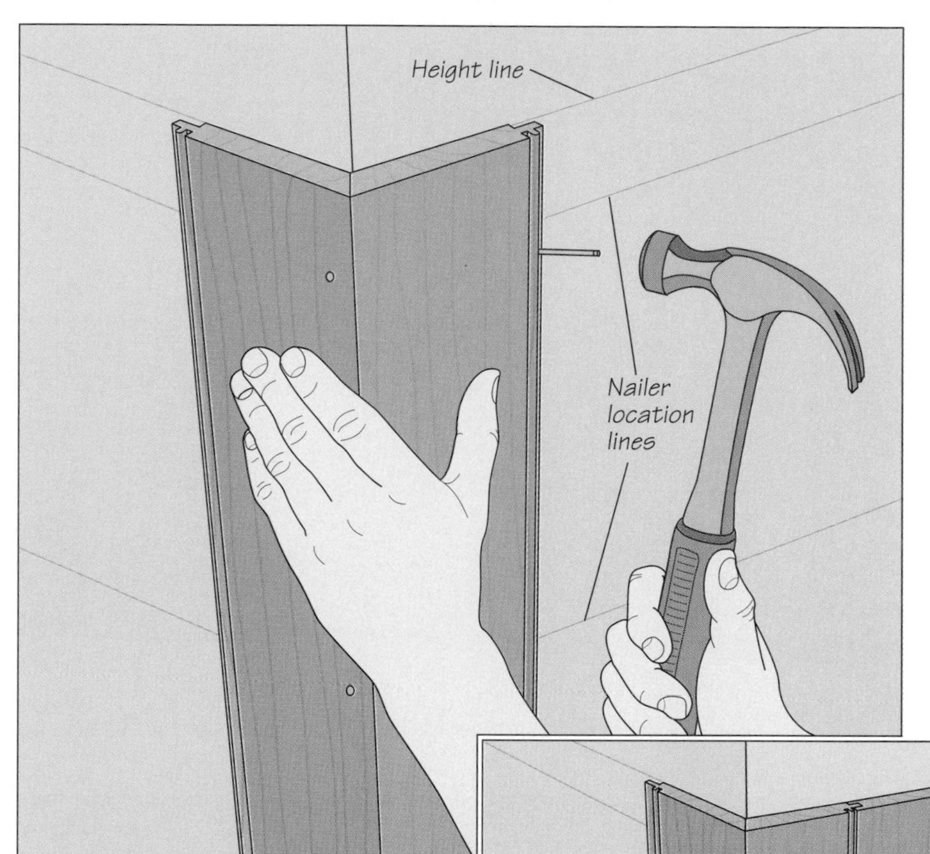

1 Installing the boards at an outside corner
Snap a chalk line around the room *(page 31)* to mark the height of the wainscoting. Make sure the line is level—the floor may not be. If you know where the nailers behind the wall are located, also snap lines for them. If you do not know where the nailers are located, install furring strips *(page 49)*. Start installing the boards at an outside corner. Determine the bevel angle you will need as you would for baseboard *(page 28)*, then make the cut along the grooved edges of two boards. To install each board, align the cut edge with the corner and drive a nail through the tongue into the wall at each nailer location line *(left)*. Nail the corner boards along the mitered edges as well. If you installed furring strips, nail the wainscoting to the strips.

2 Installing the boards along a wall
Once the two boards are installed at the outside corner, slip the groove of a new board in place. Nail the board to the wall through its tongue, making sure all the tops are level. Continue along the wall the same way, fitting grooves over tongues *(right)* and nailing the boards in place.

3 Checking for plumb
Halfway along the wall, hold a carpenter's level against the tongue of the last board you installed to check for plumb *(right)*. If the board is not perfectly vertical, taper the grooved edge of the next board with a hand plane so that it will be plumb when it is installed. Continue to the end of the wall. To fit the last board, use a log-builder's scribe to transfer the profile and angle of the adjoining wall to the face of the board *(page 42)*.

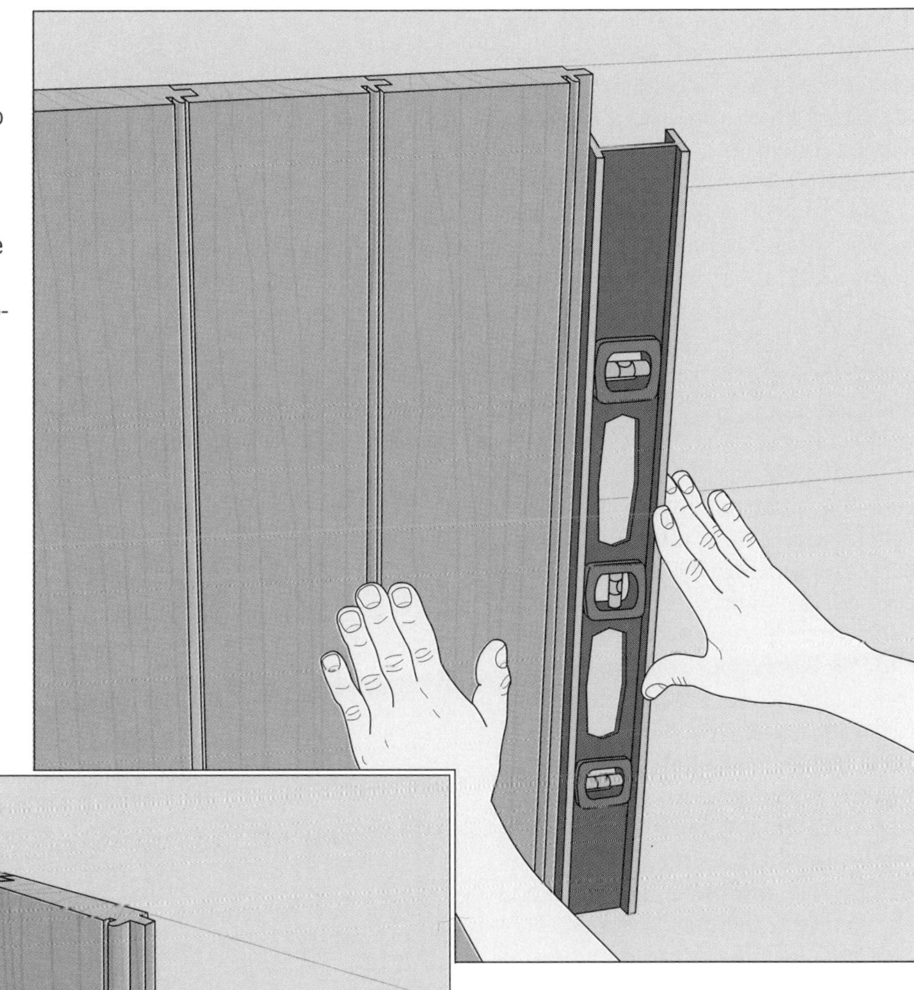

4 Paneling an adjoining wall
To install wainscoting at an inside corner, butt the grooved edge of a board against the last board you installed on the adjoining wall, then nail it in place *(left)*. Nail along the grooved edges of the boards as well. Work your way around the room until all the wainscoting is installed. Finish the job by installing baseboard *(page 22)* and adding a cap rail *(page 38)*.

FRAME-AND-PANEL WAINSCOTING

Frame-and-panel wainscoting consists of a frame of horizontal rails and vertical stiles and mullions enclosing raised panels. The frame members can be joined in a number of ways, including dowel, biscuit, mortise-and-tenon, or cope-and-stick joints. This section shows you how to cut the cope-and-stick on the shaper *(page 47)*. The panels can be raised on a table saw *(page 49)*, but specialized router bits *(page 48)* and shaper cutters *(page 109)* do a faster and cleaner job, and can shape curved profiles as well.

This type of paneling can be installed over the lower portion of a wall, like tongue-and-groove wainscoting, or it can cover an entire wall from floor to ceiling *(page 41)*. In either case, you need to determine the dimensions of the paneling and its various components. The frame pieces and panels must be sized to fit each wall exactly, and properly proportioned so they look right in the room. Be sure to consider obstructions such as doors and windows, fireplaces and ceiling beams. Before cutting any wood, make a scale drawing of the room and experiment with different designs.

When installing frame-and-panel wainscoting, it is best to work on one wall at a time, preparing and installing the frame pieces and panels together. The wainscot can then be glued up and installed as a unit, or built up on the wall piece by piece. If you follow the latter method, you will be able to compensate for any mistakes as you go along.

The cope-and-stick joint shown in the photo at right is an easy-to-cut alternative to the mortise-and-tenon traditionally used in frame-and-panel wainscoting. It also adds a decorative touch: The router bit that cuts the grooves for the panel and tongues in the stiles and rails also carves a molding along the inside edges of the frame.

A GALLERY OF RAISED PANEL PROFILES

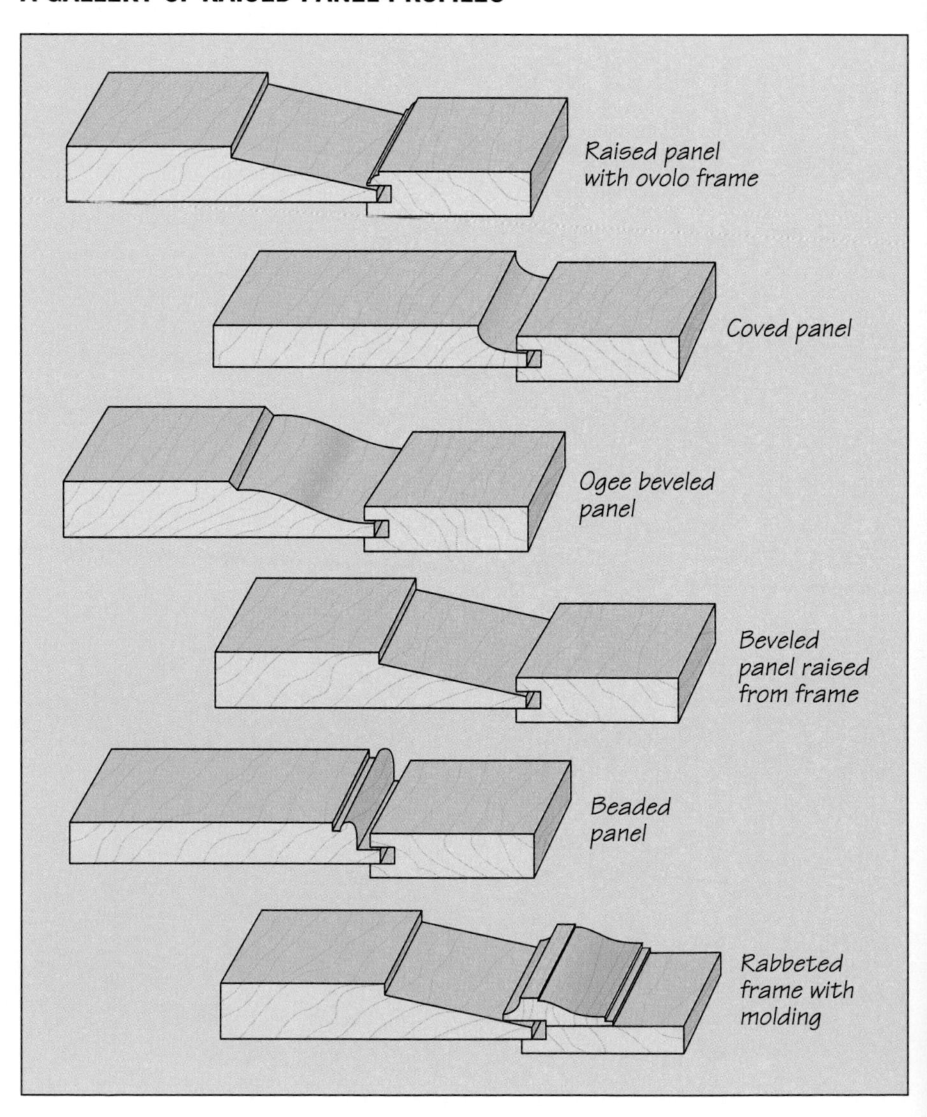

MAKING A COPE-AND-STICK FRAME

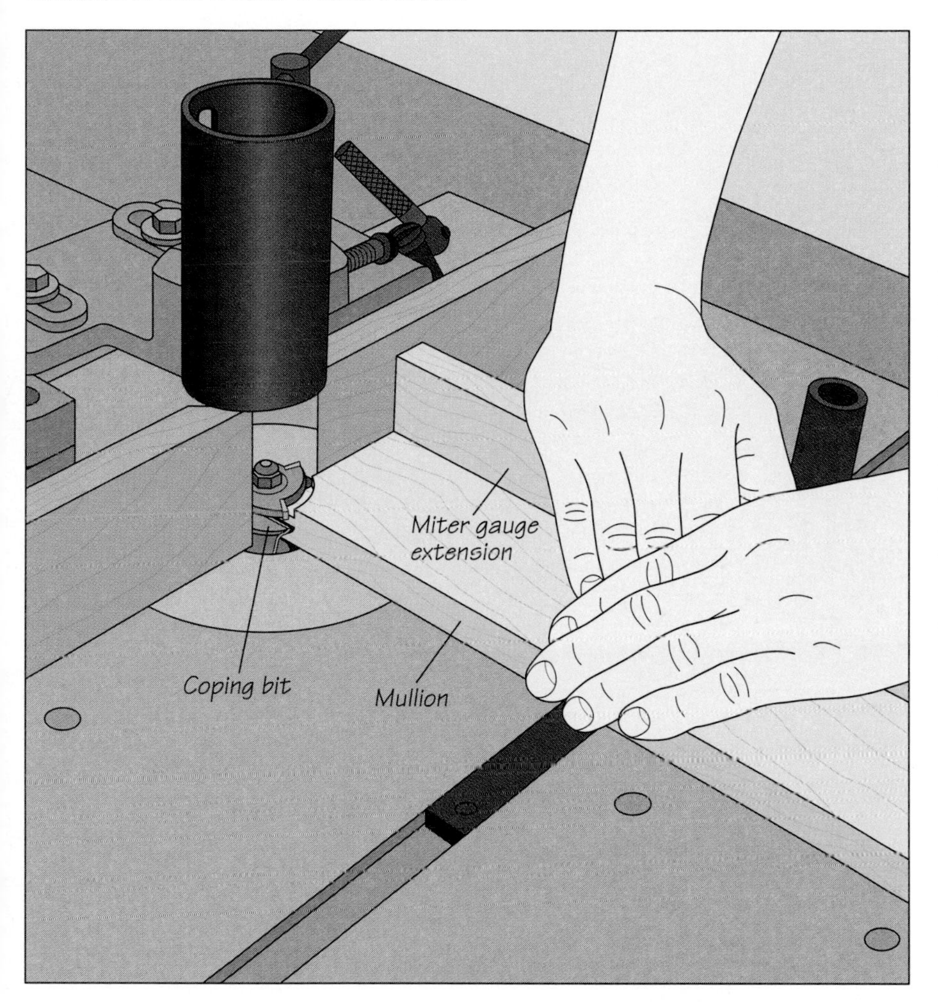

1 Cutting tongues in the rails and mullions

Saw the rails, mullions, and stiles to length. To join the pieces of the frame with cope-and-stick joints, start by cutting tongues at the ends of the rails and mullions, as shown at left. Install a piloted coping bit in your shaper and adjust the cutting depth by butting the end of a rail against the cutter and setting the top of the uppermost cutter slightly above the workpiece. Position the fence parallel to the miter gauge slot and in line with the edge of the bit pilot. For added stability, screw a board to the miter gauge as an extension. Feed the stock face down with the gauge, holding the edge of the workpiece against the extension and the end against the fence *(left)*. (Although a shaper is used here, a table-mounted router can also be used to cut cope-and-stick joints.)

2 Adjusting the sticking bit

Replace the coping bit with a piloted sticking bit. To adjust the cutting depth, butt the tongue at the end of a rail or muntin against the bit and set one of the groove-cutting teeth level with the tongue *(below)*. Align the fence with the edge of the bit pilot.

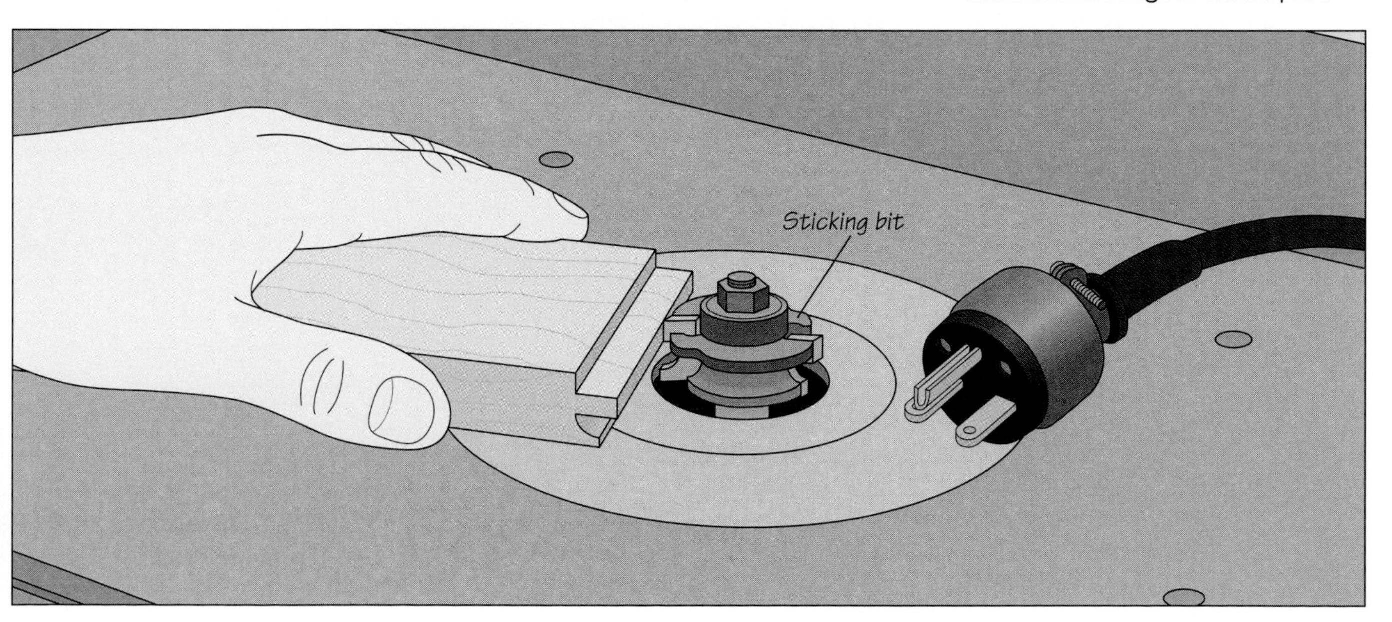

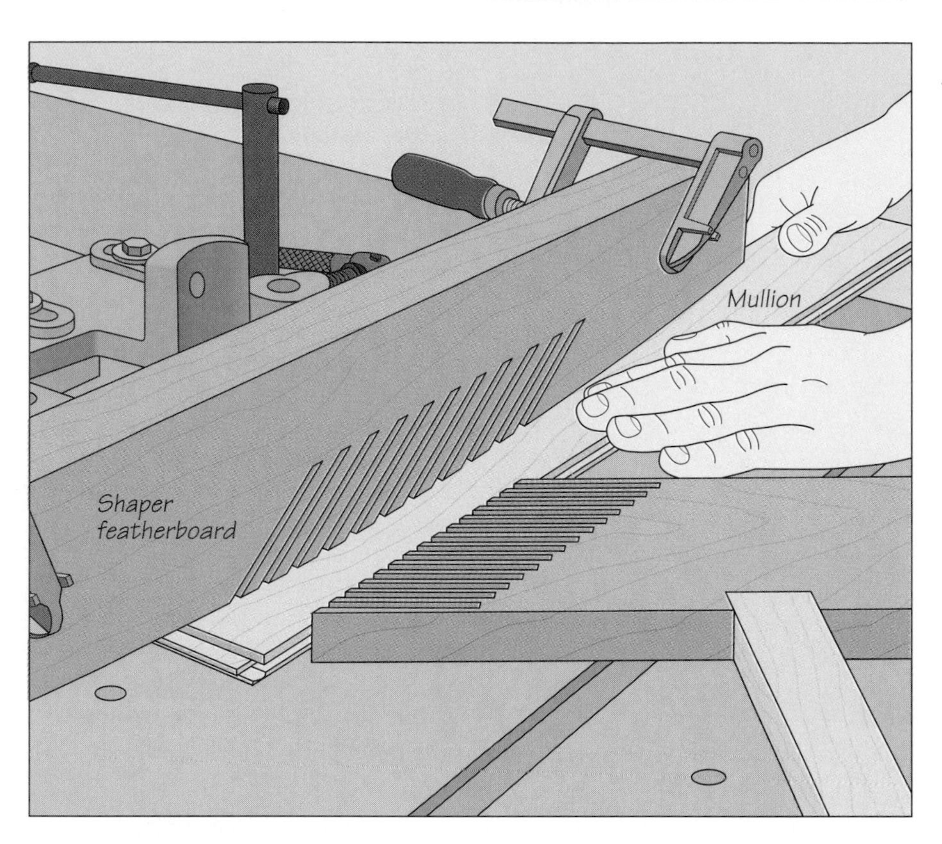

3 Cutting the grooves and decorative profile

Use two featherboards to secure the stock during the cuts. Clamp a standard featherboard to the shaper table opposite the bit; secure a support board at a 90° angle to the featherboard. Clamp a shaper featherboard to the fence. Make this featherboard on the band saw by curving the bottom edge of a 2-by-4 and cutting a series of angled slots into the edge. Cut the grooves and decorative profile along the inside edges of the stiles and rails, and along both edges of all the mullions. Make each pass with the stock outside-face down, pressing the workpiece against the fence *(left)*. Use a push stick to complete the pass.

RAISING THE PANELS

Raising a panel on the router table

Cut each panel to fit within its frame, adding ¼ inch on all sides for the grooves. Install a piloted panel-raising bit in your router and mount the tool in a table. Set the depth of cut at ⅛ inch and clamp two featherboards to the fence, one on each side of the bit. To raise the panel, feed the board across the table outside-face down, keeping the workpiece flush against the fence and your hands clear of the cutter *(right)*. To minimize tearout, cut into the end grain of the panel first, shaping the two ends of the panel before the sides. Test fit the panel in the frame grooves and make as many passes as you need, increasing the cutting depth no more than ⅛ inch at a time.

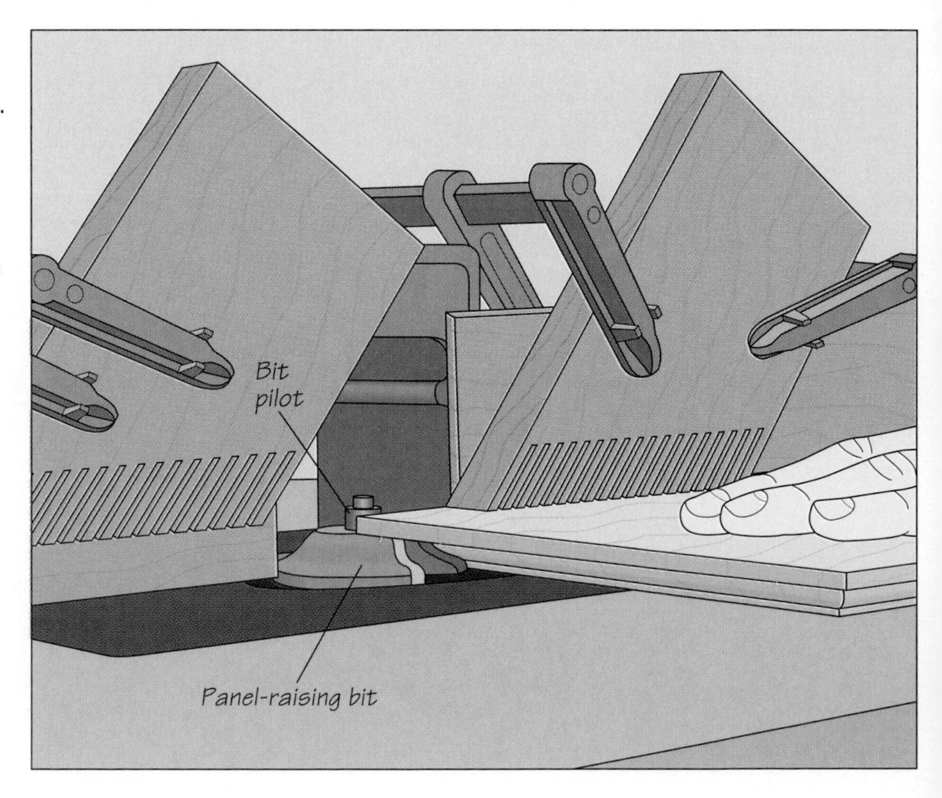

Raising a panel on the table saw
To set the proper blade angle for raising a panel on the table saw, mark a cutting line on the panel: Draw a ¼-inch square at the bottom corner, then mark a line from the front face of the panel through the inside corner of the square to a point on the bottom edge ⅛ inch from the back face *(inset)*. Install an auxiliary wood fence, set the panel against it, and adjust the angle of the blade until it aligns with the cutting line. Raise the blade until one tooth just protrudes beyond the front face of the panel. Make a cut in one end of the panel and test-fit the cut in a groove. If the panel sits less than ¼ inch deep, move the fence a little closer to the blade and make another pass. To minimize tearout, bevel the ends of the panel first, then the sides *(right)*.

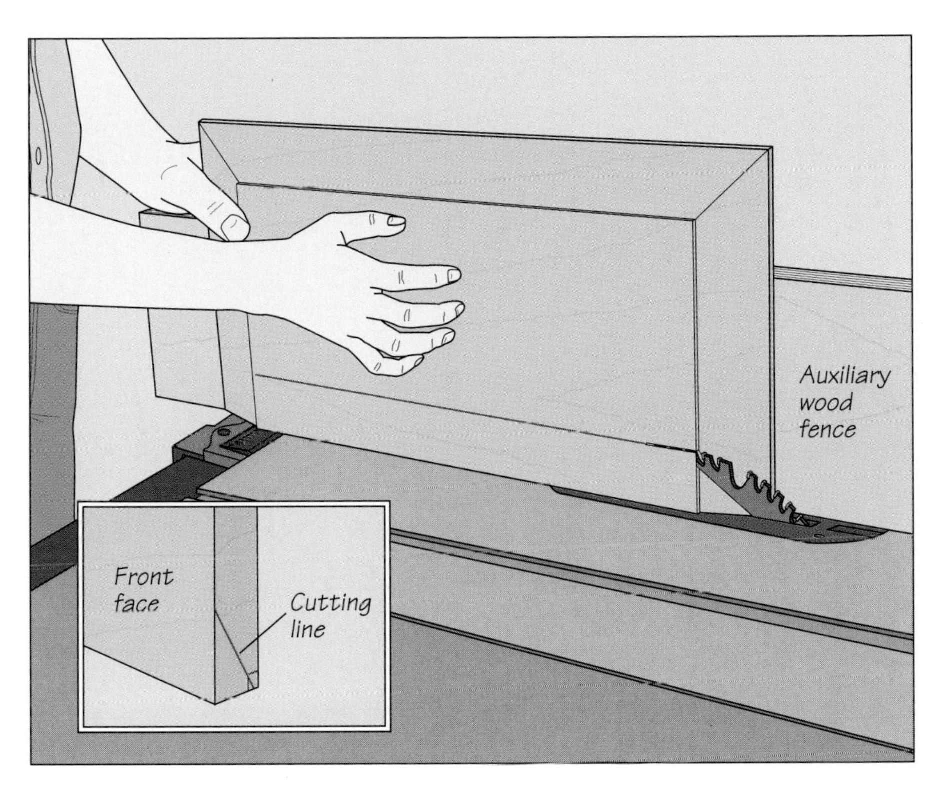

INSTALLING FRAME-AND-PANEL WAINSCOTING

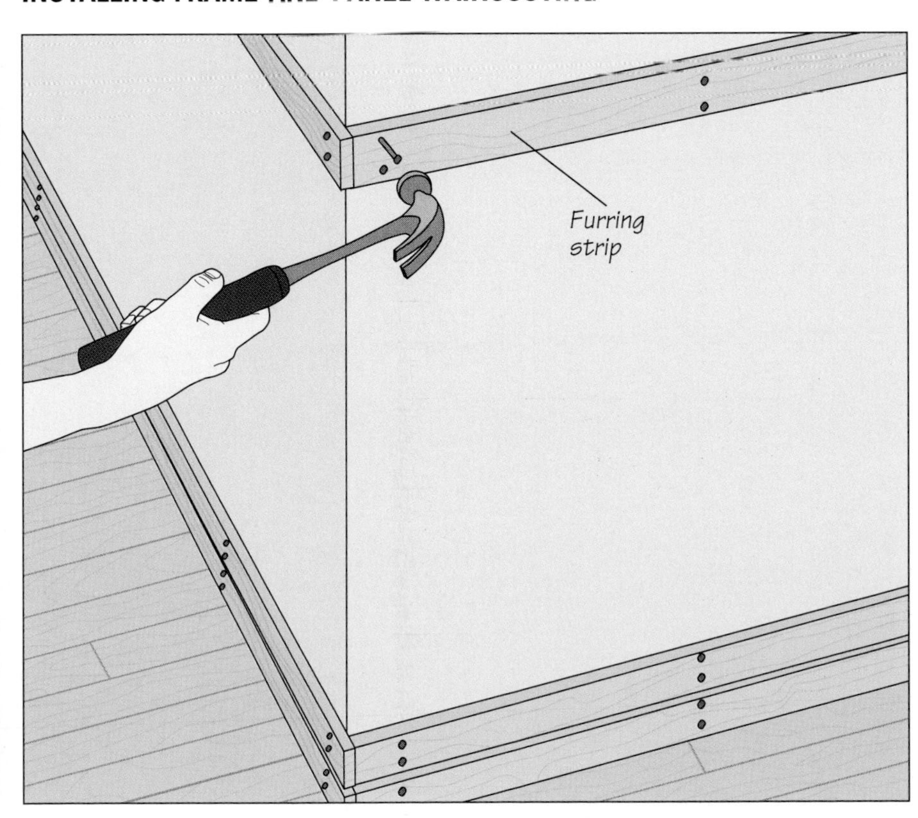

1 Installing furring strips
Like tongue-and-groove paneling, frame-and-panel wainscoting can be fastened to nailers behind the wall *(page 44)*. If you do not know where the nailers are located or whether they exist, you will have to anchor the paneling to furring strips. Snap two chalk lines on the wall to help you install the strips. For frame-and-panel wainscoting, one strip should be level with the top rail, typically 36 inches off the floor. Locate the second chalk line a few inches above the floor. Saw the furring strips from 1-by-3 stock; cut one for the top rail and two for the bottom rail. Determine the location of the studs *(page 32)* and nail each furring strip in place *(left)*, driving two nails at each stud.

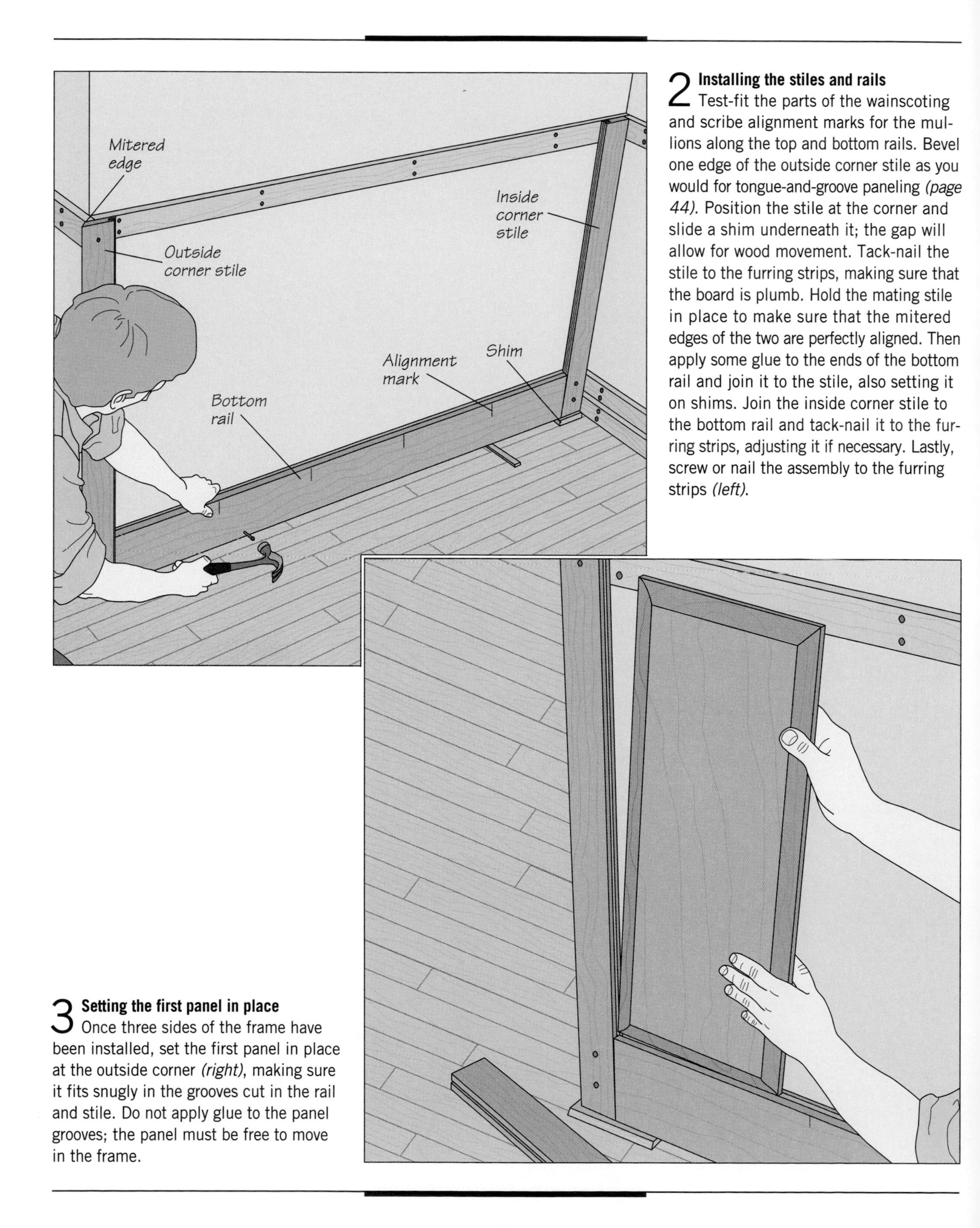

2 Installing the stiles and rails
Test-fit the parts of the wainscoting and scribe alignment marks for the mullions along the top and bottom rails. Bevel one edge of the outside corner stile as you would for tongue-and-groove paneling *(page 44)*. Position the stile at the corner and slide a shim underneath it; the gap will allow for wood movement. Tack-nail the stile to the furring strips, making sure that the board is plumb. Hold the mating stile in place to make sure that the mitered edges of the two are perfectly aligned. Then apply some glue to the ends of the bottom rail and join it to the stile, also setting it on shims. Join the inside corner stile to the bottom rail and tack-nail it to the furring strips, adjusting it if necessary. Lastly, screw or nail the assembly to the furring strips *(left)*.

3 Setting the first panel in place
Once three sides of the frame have been installed, set the first panel in place at the outside corner *(right)*, making sure it fits snugly in the grooves cut in the rail and stile. Do not apply glue to the panel grooves; the panel must be free to move in the frame.

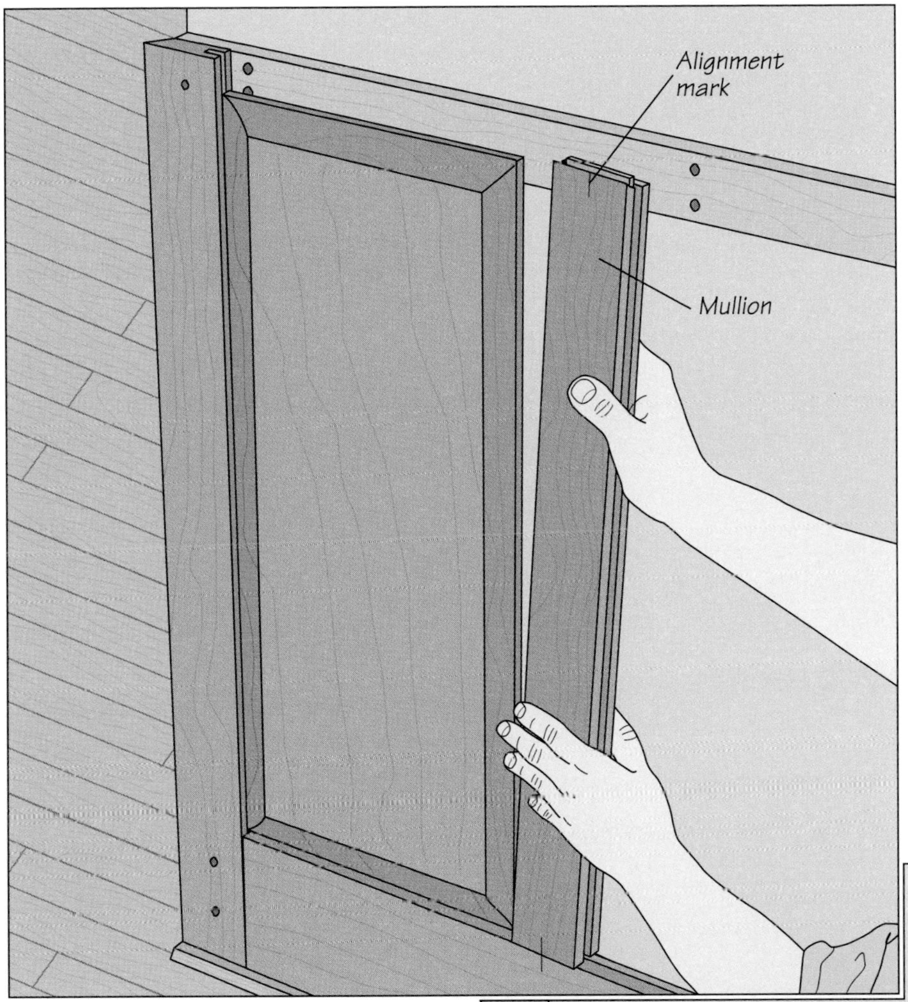

4 Installing the first mullion

Apply glue to the bottom end of the mullion and the groove in the bottom rail, and set the mullion in place *(left)*. Make sure the panel sits in the grooves cut in the mullion's edge. Continue installing panels and mullions until you reach the stile at the inside corner. Slip the last panel between the mullion and the stile.

5 Installing the top rail

Apply glue to the top ends of all the mullions and fit the top rail in position *(right)*, making sure the top ends of the panels, mullions, and stiles all fit snugly in the groove in the top rail. An extra set of hands will make the job easier. Once the top rail is in place, you can install baseboard molding *(page 24)* and a cap rail *(page 38)*. Proceed to the next wall and work your way around the room, using butt joints at inside corners and mitering outside corners.

PANELED CEILINGS

Paneled ceilings, also known as coffered ceilings, are an adaptation of frame-and-panel techniques normally used to decorate walls. In conjunction with frame-and-panel wainscoting *(page 46)*, a paneled ceiling can add depth and warmth to a den or study.

The illustration below shows how a paneled ceiling is installed. Start with a structural framework of 2-by-4s anchored to the ceiling joists. Cover the framework with 1-by-4 hardwood stock, such as oak or birch, then make box-like facings to fit inside the framework, creating a grid of boxes. Finally, set a veneered plywood panel into each box and install a frame of molding to hold the panel in place.

As with full-wall paneling *(page 41)*, the size of the panels should be proportional to the dimensions of the room. A ceiling of small panels in a large room looks cluttered, while the opposite can appear too heavy. Determine the panel size by making a scale drawing of the ceiling and experimenting with different dimensions. A panel size between 20 and 26 inches is typical for medium-sized rooms.

You can try variations on this basic design. The panel boxes may be left white for contrast or crown molding can be used in place of quarter-round molding—although this would involve cutting coped joints at inside corners. If you decide to use stain, it is a good idea to apply the finish before installing the ceiling, for working overhead on a ceiling full of crevices can prove tiring.

AN ANATOMY OF A PANELED CEILING

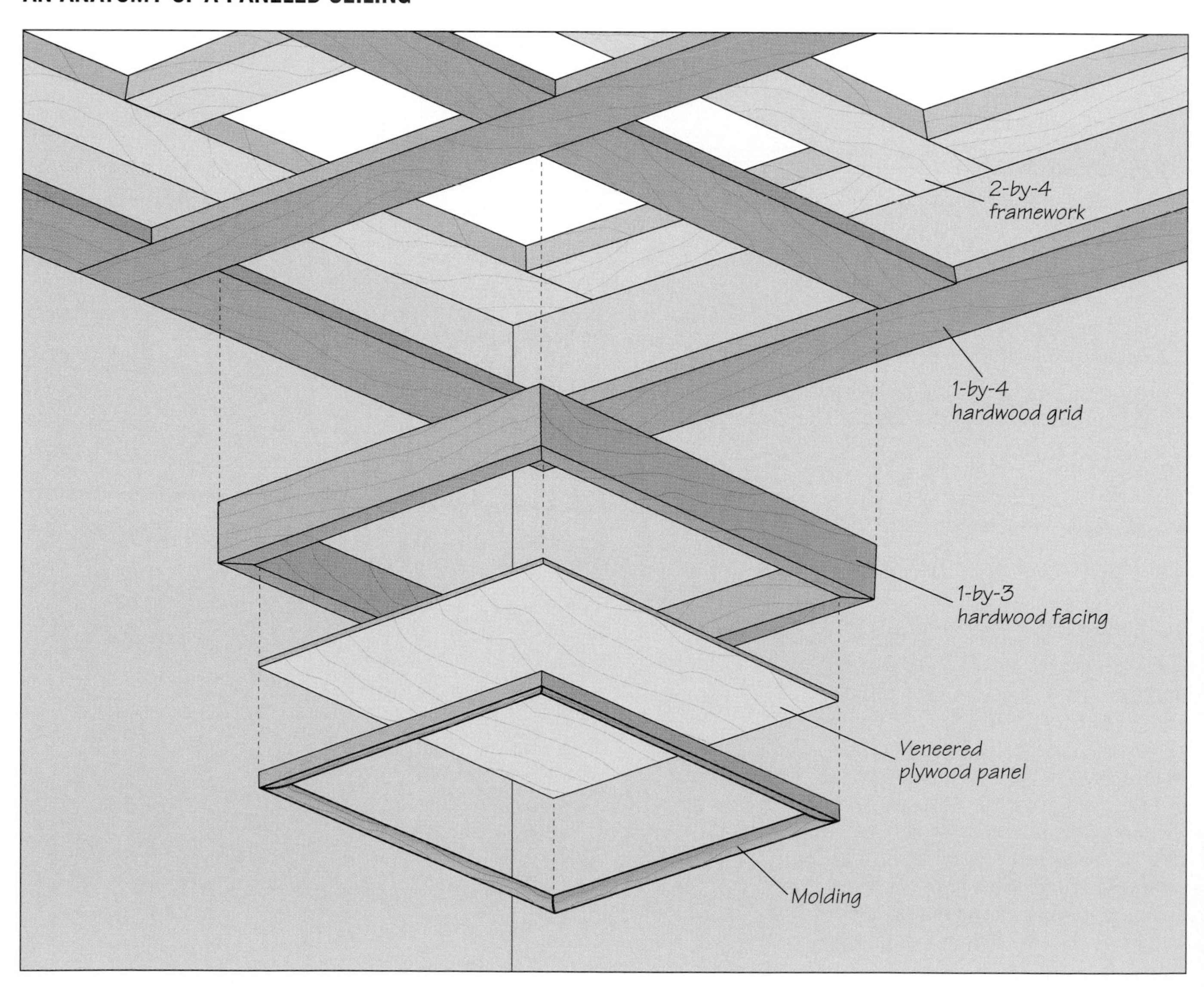

INSTALLING A PANELED CEILING

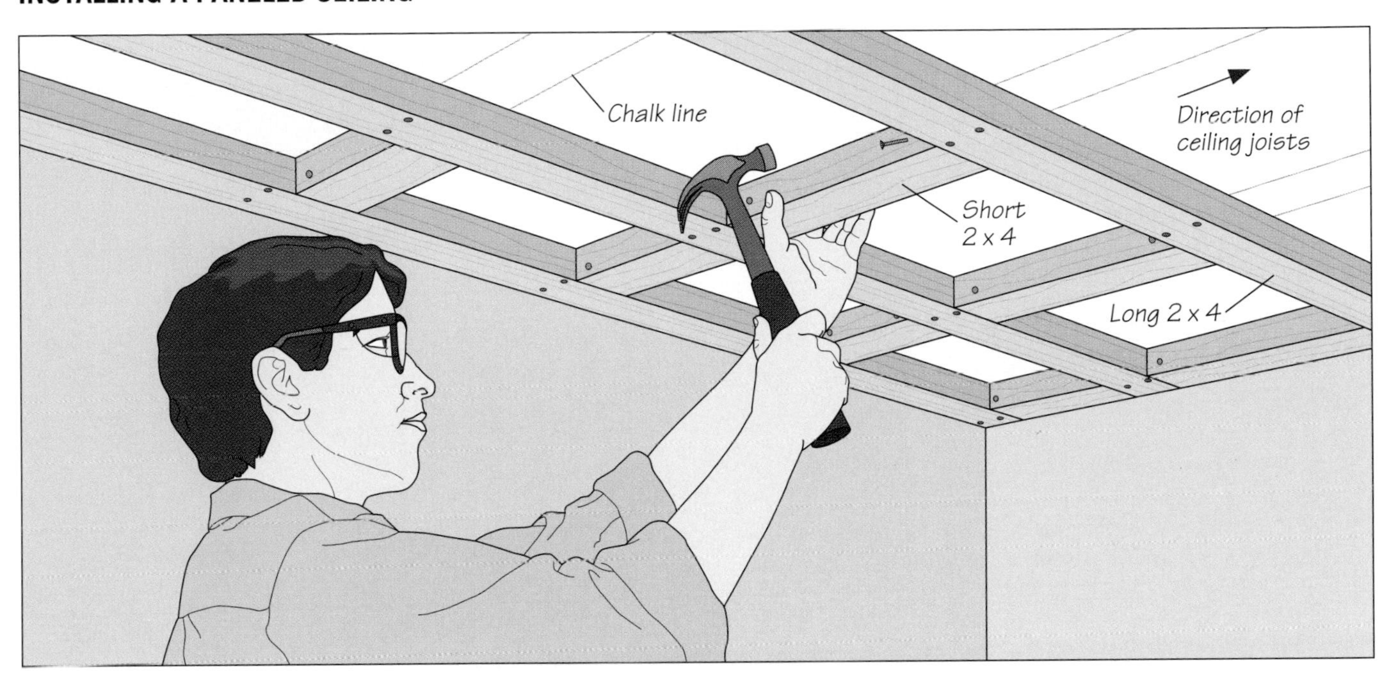

1 Installing the 2-by-4 framework
Snap a grid of chalk lines on the ceiling based on the size of your panels. Start from the center of two opposite walls so that any small panels will be located around the edge of the ceiling. Remember to account for the width of the 2-by-4s as you lay out the chalk lines. Determine the direction of the ceiling joists with a stud finder, then install the framework. Use long 2-by-4s to span the ceiling perpendicular to the joists; align the pieces with the chalk lines and fasten them to the joists with flooring screws. Use shorter 2-by-4s to fit between the long boards and toe-nail them in place *(above)*.

2 Installing the hardwood grid
Once the 2-by-4 framework is installed, use a finish nailer to mount the 1-by-4 hardwood grid *(left)*. For added rigidity, install the long 1-by-4s perpendicular to the long 2-by-4s. Although the nailing can be done by hand, a finish nailer makes the work go much faster.

3 Installing the facing
Within each frame, install four facing pieces of 1-by-3 hardwood stock that matches the wood you used for the grid. For a more decorative effect, rip the pieces so they extend below the grid by ¼ inch. Fit the pieces at the corners with 45° bevel cuts *(right)*, then nail them to the 2-by-4 framework.

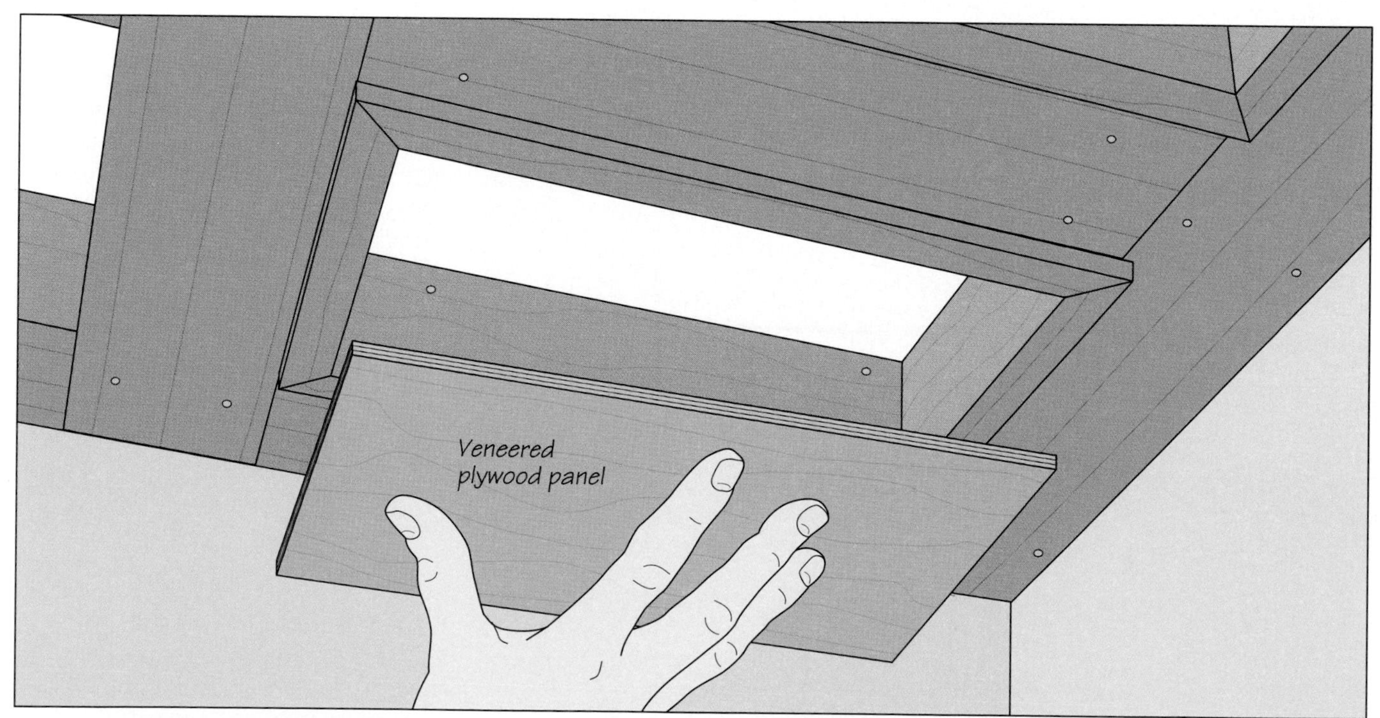

4 Gluing the panels in place
Once all the facing is installed, cut panels from veneered plywood to fit within the frames. Apply construction adhesive to the underside of each panel and press it in place against the ceiling *(above)*. The adhesive will hold the panel until you install the shoe molding *(step 5)*. The molding will supply pressure needed for a firm glue bond.

5 Installing the quarter-round molding
Secure the panels with quarter-round molding installed around the inside of each facing box. Cut the molding to length, joining the pieces at the corners with 45° miters. Nail the molding to the facing with a finish nailer *(left)*.

6 Installing crown molding
To finish the ceiling, install crown molding around its perimeter. Nail it to the framework and wall as you would on a plain ceiling *(page 26)*, using coped joints at inside corners and miters at outside corners.

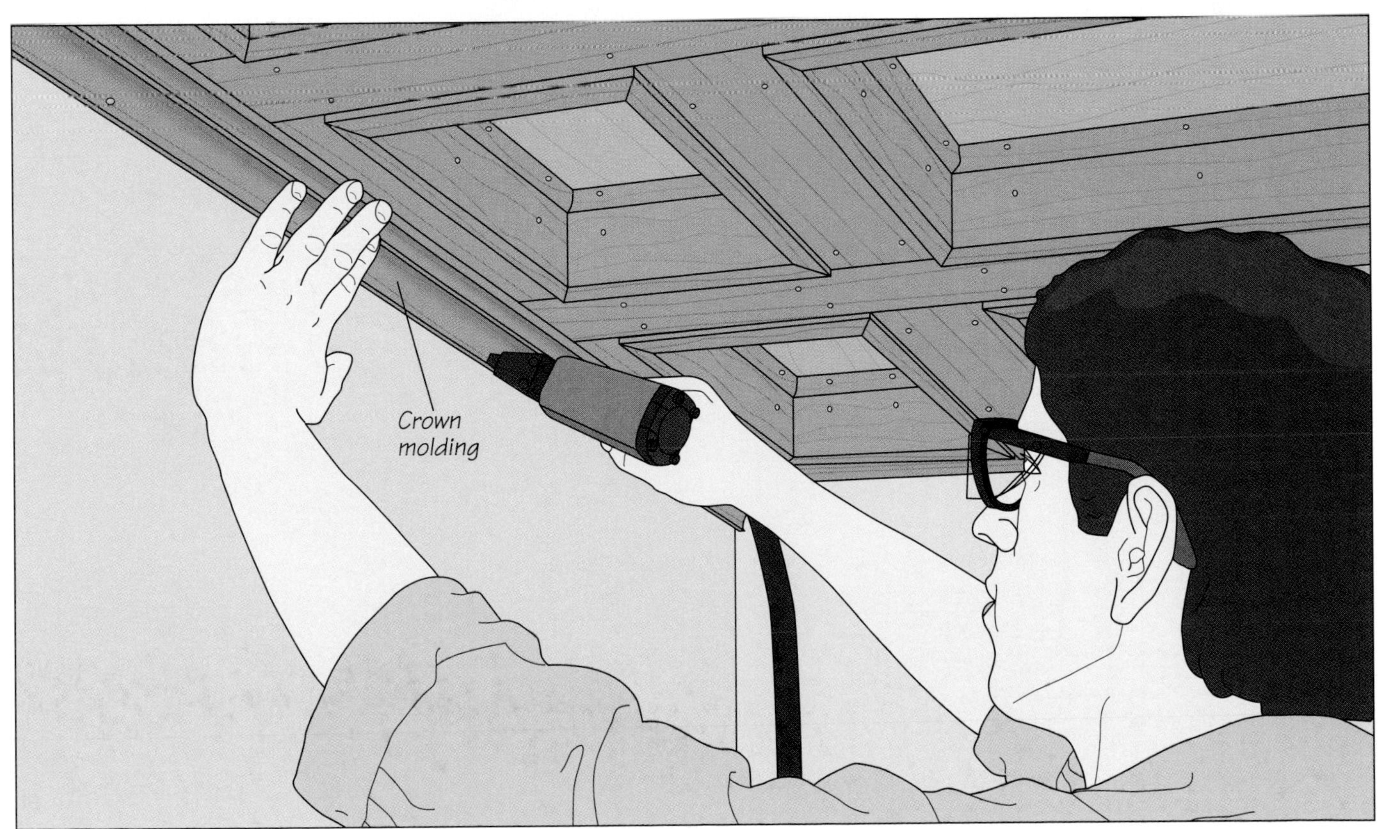

WINDOWS

A good window lets all the light and beauty of the outdoors in while keeping the elements out. Although they do this in many ways, reflecting a wide range of styles, windows consist of two basic parts: the frame and the sash. The former is like a door jamb, and serves much the same purpose: It is fixed to the studs when the window is installed. The sash actually holds the pane of glass. The windows covered in this chapter feature the popular double-hung sash. As shown on page 59, they contain two sashes, both of which can slide up and down.

Once a window is installed *(page 59)*, gaps remain between the frame and the surrounding framing of studs and headers. Just as sills, stools, aprons, and casings are installed on the outside of a window to complement the exterior trim of a house, the window needs to be "cased," or framed on the inside as well. Finish carpenters use two basic methods for this task: picture-frame and stool-and-apron.

A mitered "return" is glued onto the apron below a window sill, or stool, to hide the end grain of the apron. Stool-and-apron is a traditional method of casing a window, often made to match the room's interior trim. The molded casing hides the gaps between the window jambs and the wall.

Picture-frame casing *(page 61)* consists of four pieces of molding: two side casings, a head casing, and a sill casing, all joined at the corners with 45° miters. Stool-and-apron casing *(page 69)* also features two side casings, and may include decorative corner blocks, known as rosettes *(page 73)*.

In the example shown in this chapter, the head casing is butted against the side casings, but these joints can also be mitered. The most recognizable element of stool-and-apron casing, however, is the stool, or sill, installed at the bottom, which juts out from the window.

While many home builders, particularly those in colder climates, opt for the precision and insulation of factory-made windows, elegant, high-quality windows can be made in the shop with specialized sash cutters *(page 75)*. Since these windows require thicker stock than most cabinet work, a shaper is the stationary power tool of choice for making them.

The final step in making a custom-fit window is installing the glass-stop molding, thin strips of shaped wood that hold the panes of glass in place. Shaped on a router table and ripped to width on the table saw, the molding strips are joined at the corners with miters and nailed to the window sash.

BASIC WINDOW TRIM STYLES

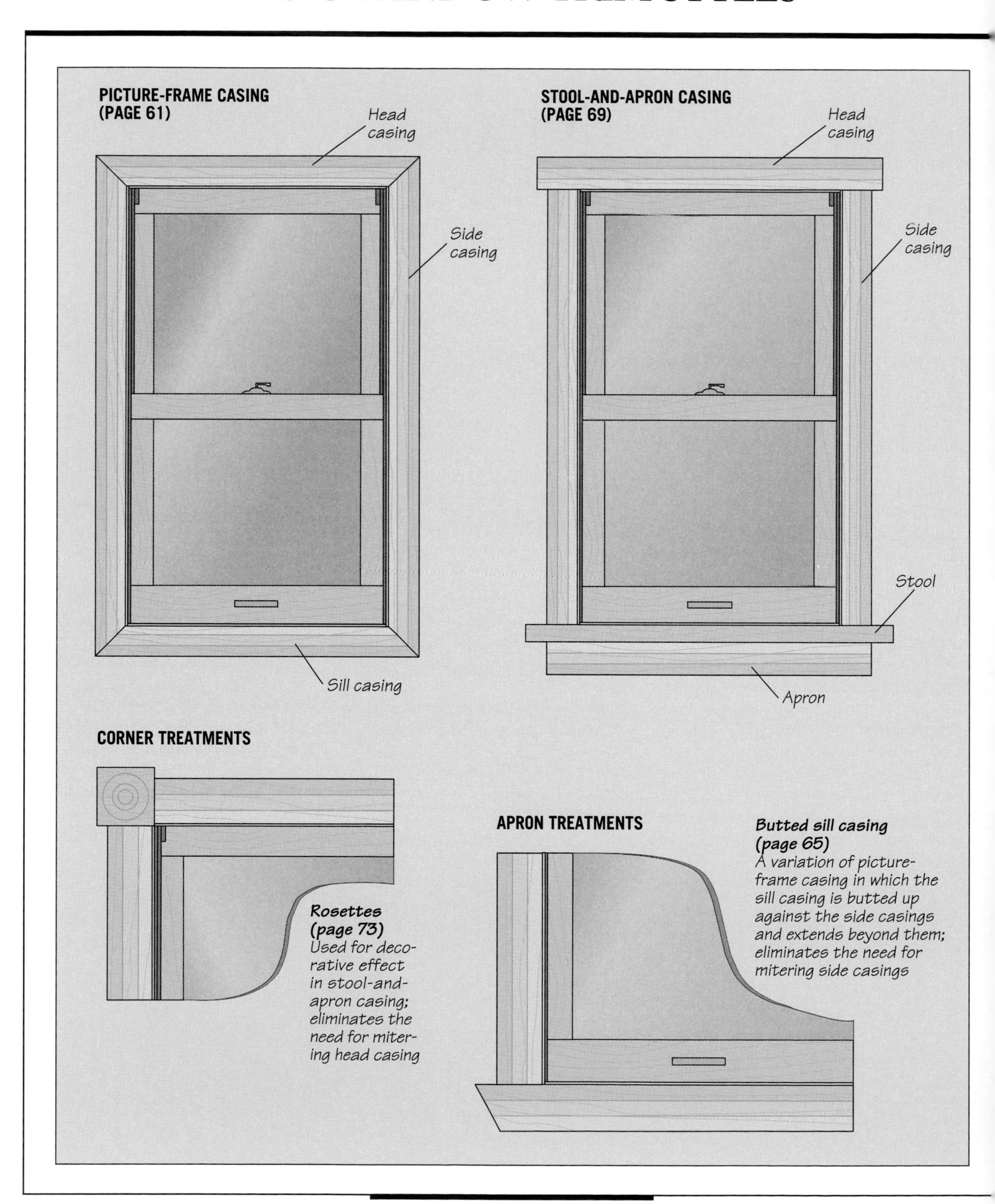

INSTALLING WINDOWS

Although fitting a window in place may appear to be a daunting task, it can in fact be a straightforward operation. With the pre-hung window shown below, which already includes the window jambs attached to the sashes, all you need is a hammer or a screwdriver and a level. Whether your windows are factory- or shop-made, they will be installed in the same fashion. The jambs are nailed into the rough opening in the wall, then insulated and dressed with interior trim. Sometimes, a jamb extension *(page 61)* is installed on the inside to bring the window flush with the interior wall.

A window should be about ½-inch smaller on all sides than its rough opening. Since rough openings are seldom square, level, or plumb, this will make the window easier to fit and shim, while leaving some space around the window for insulation. Remember not to drive the shims in too far or you may risk bowing the window. Test the window to make sure that it slides smoothly before nailing it in place.

A double-hung window is positioned in its rough opening from the outside. To help hold the window in place until it can be adjusted and secured from the inside, it will be nailed or wedged temporarily to the frame of furring strips around the opening.

INSTALLING A DOUBLE-HUNG SASH WINDOW

1 Leveling and centering the window
Position the window in its rough opening *(photo, above)* and temporarily tack or wedge it in place. To help check the window for level as you go, clamp a carpenter's level to the underside of the head jamb. Insert shims between the side jambs and studs at the top of the rough opening. (Shims are wooden wedges usually sold in bundles at hardware stores and lumberyards.) Then, holding up one corner of the window, slip a shim between the window horn and the rough sill *(left)*. Repeat on the other side of the window. Add shims between the side jambs and the studs at the middle and bottom of the window. Use as many shims as you need to center the window in its opening while keeping the window level. (To install more than one window at the same height in a room, make a mark on the stud at a set distance from the header and shim all the windows to the mark.)

2 Nailing the window in place
Once the window is level, fasten it to the wall framing. Drive a finishing nail through the side jambs and shims into the studs at each shim location *(above)*. Cut the shims flush with the window jambs using a utility knife.

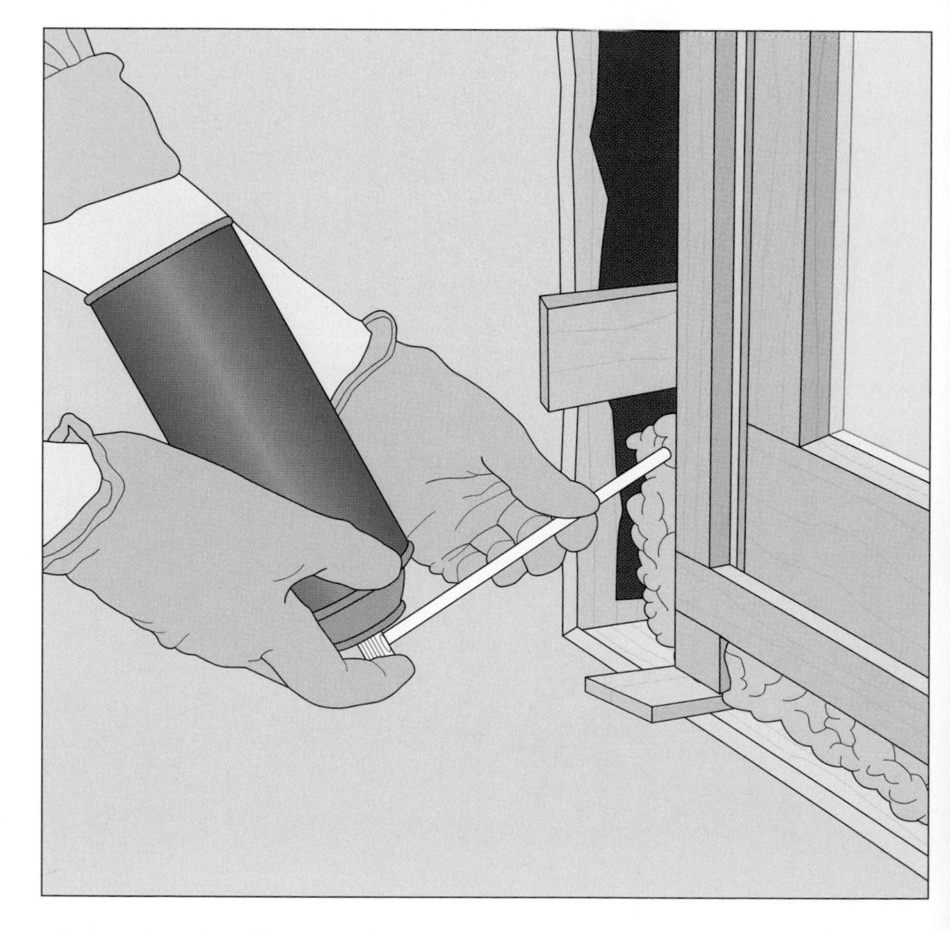

3 Insulating the window
Once a window is nailed in its rough opening and before installing the casing, it is a good idea to fill the hollow spaces between the window jambs and the wall studs with insulation—particularly if you live in a cold climate. You can use fiberglass insulation or a low-expanding foam as shown here. Fiberglass works well for large gaps, such as the space between the head jamb and the header or the space between the two sills. Foam insulation is ideal for thinner spaces, but use it sparingly; too much of it may cause the jambs to bow inward.

PICTURE-FRAME CASING

Picture-frame casing comprises four pieces of molding mitered at 45° that frame a window in much the same way as a picture is framed. Commercial picture-frame casing is available in a number of profiles. It can also be milled on a table saw fitted with a molding head using the same procedures to produce chair rail, custom baseboard, or any other molding *(back endpaper)*.

In order to nail picture-frame casing in place, the front edges of the window jambs need to be flush with the interior wall. If the jambs are set more than ¼ inch back from the drywall, you will need to build and install a jamb extension *(below)*. As the casing will hide the extension, the joinery used to attach the extension pieces together can be as simple as a butt joint.

A shop-made gauge is used to mark out the narrow portion of the window jambs that will not be covered with casing. This exposed portion of the jambs—anywhere from ⅛ to 5⁄16 inch wide—is called the "reveal." It both enhances the visual effect of the casing and makes it easier to install. To make the jig, see page 63.

INSTALLING A JAMB EXTENSION

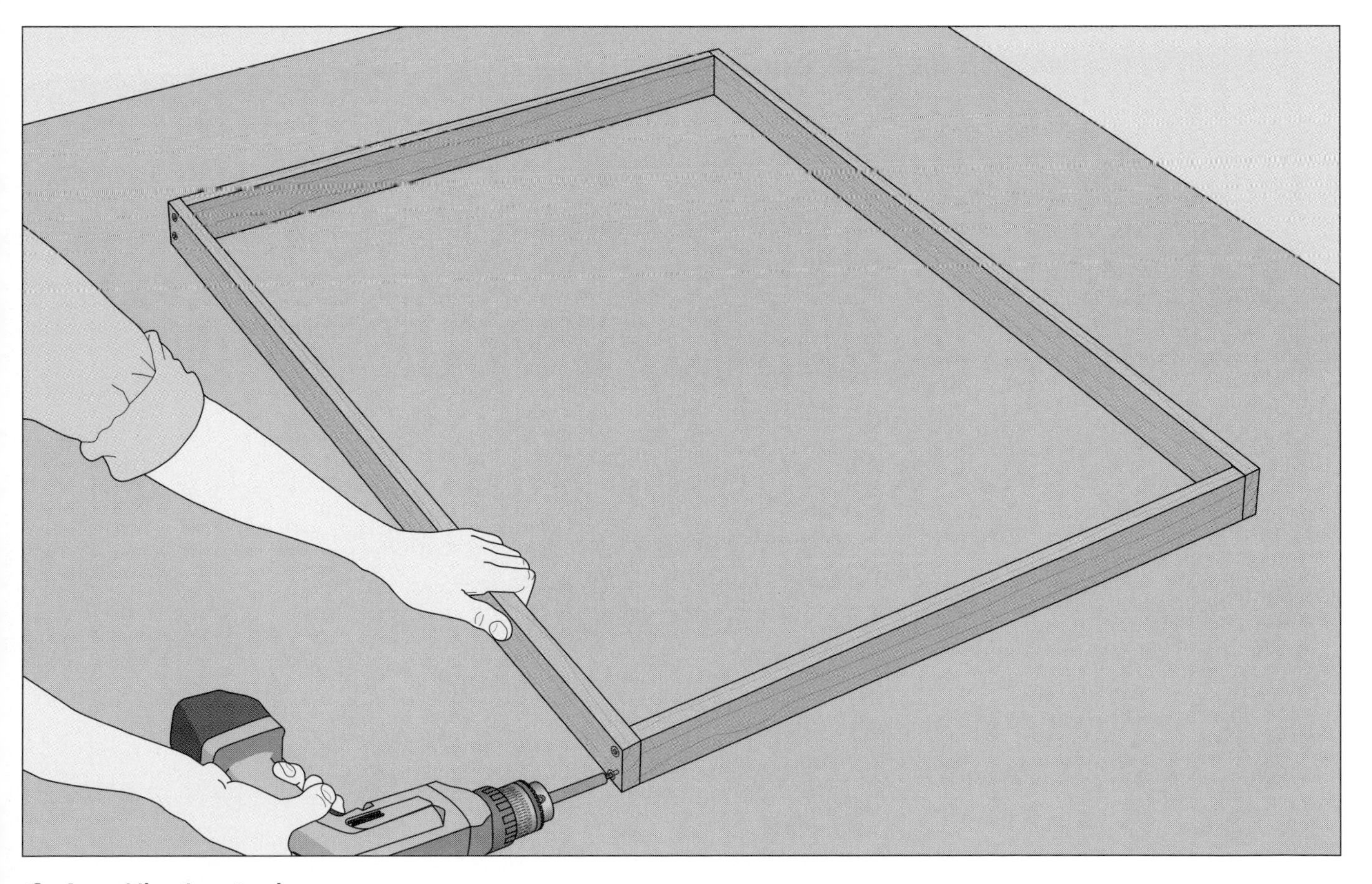

1 Assembling the extension

Measure the distance between the front edges of the jamb and the inside wall. Then rip your extension stock to this width from wood the same thickness as the window jamb. Cut the pieces to size to make a frame that will fit the inside faces of the window jambs with a slight reveal. You can install the extension pieces one by one, or nail or screw them together into a unit *(above)*.

2 Installing the extension
Fasten the jamb extension in place as you did to install the window *(page 59)*. Position the extension over the jamb, using shims to ensure that it is level *(right)*. Then nail the extension to the wall studs through the shims. Add insulation in the empty spaces around the extension.

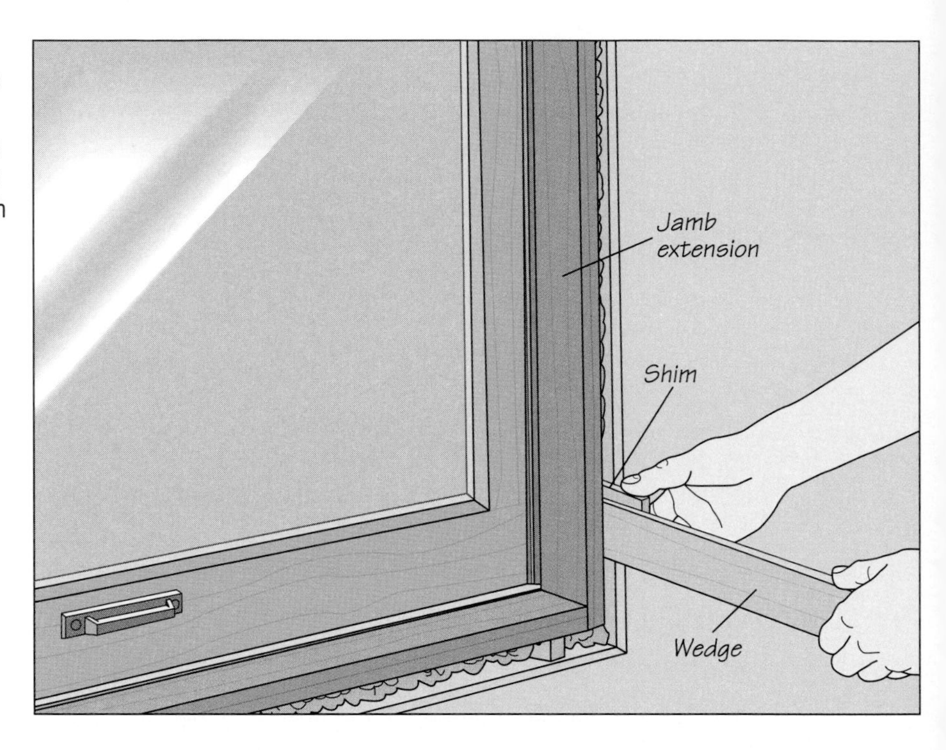

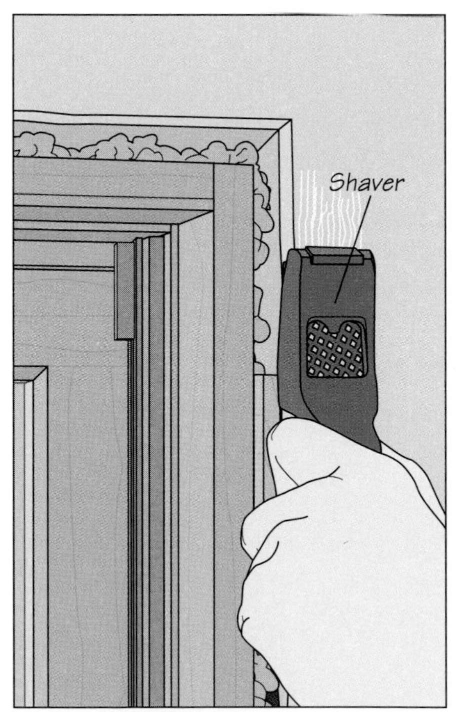

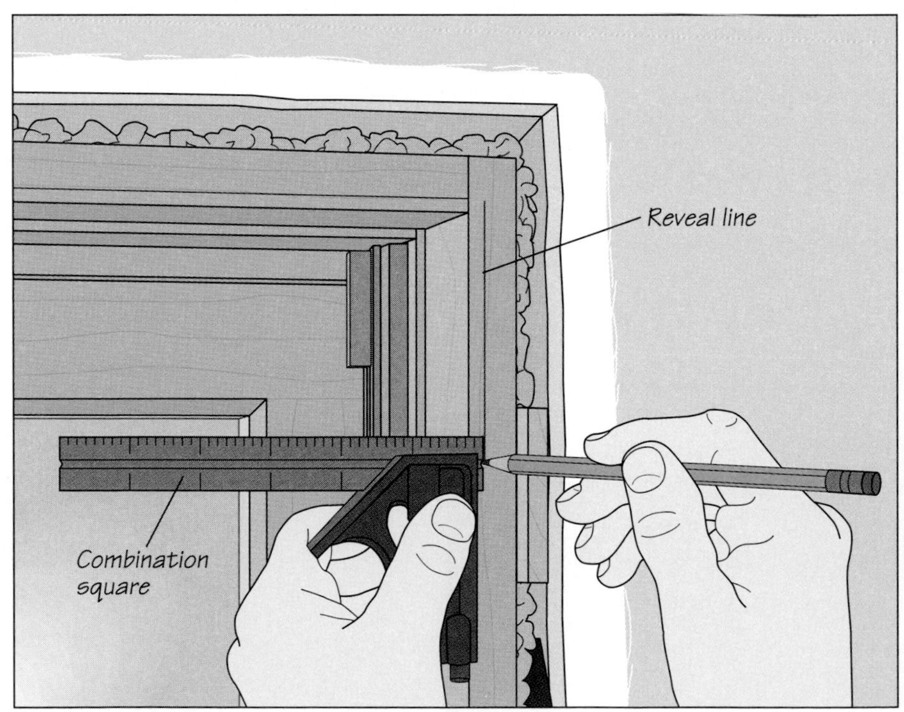

3 Marking the reveal
If the jamb extension is slightly proud of the interior wall, plane it down. If it is set back from the wall, use a rasp or a shaver *(above, left)* to cut the drywall down until it is flush with the extension. Avoid cutting into any part of the wall that will not be covered by the casing. Next, mark the reveal around the jamb or the extension. Adjust a combination square to the desired reveal—typically between ⅛ and 5/16 inch. Then, starting at the head jamb, butt the square's handle against the inside face of the jamb. With a pencil flush against the blade, slide the handle down the jamb to mark the reveal line *(above, right)*.

BUILD IT YOURSELF

A REVEAL GAUGE

The shop-made jig at right makes it easy to mark the reveal for casing around window jambs. To make the gauge, cut a square piece of ¾-inch plywood or hardwood, then saw a different-sized rabbet in each of the four edges. Each rabbet width should correspond to a typical reveal width—in this case, ⅛ inch, 3/16 inch, ¼ inch, and 5/16 inch. Mark the widths on each side. Do not make your reveals too wide, otherwise you will have to drive the nails near the edge of the casing, which will risk splitting it. For a ¾-inch jamb, a reveal of ¼ inch is about right. To use the reveal gauge, butt the appropriate rabbet against the jamb and slide it down the jamb with a pencil *(page 61)*.

Reveal gauge
¾" x 4" x 4"

INSTALLING PICTURE-FRAME CASING

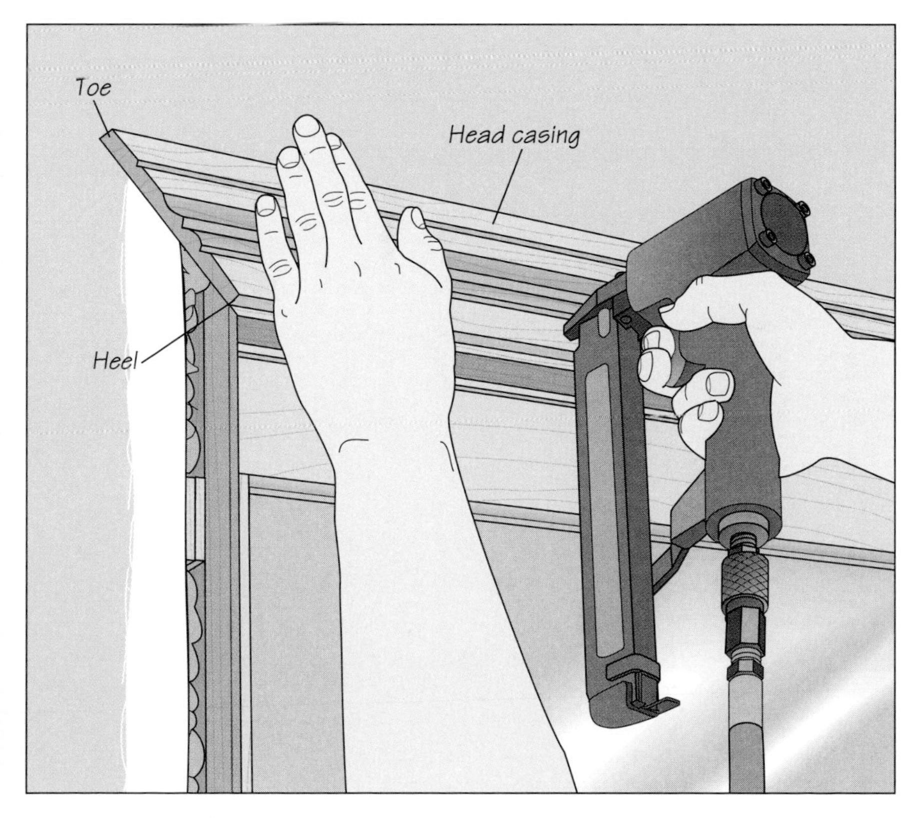

1 Installing the head casing
To determine the length of the head casing, measure the distance between the window jambs and add twice the reveal to your measurement. Miter both ends of the head casing at 45° so that the distance between the heels of the miters equals your result. Then, aligning the bottom edge of the casing with the reveal line, fasten the head casing in place with a hammer or finish nailer *(left)*. Space a pair of nails every 6 inches, driving one into the jamb and the other directly above it through the wall and into the header.

2 Installing the side casing
Determine the length of the side casing pieces and miter their ends as you did the head casing. Set the pieces in place; if either miter joint fits poorly, correct the fit as described starting on page 66. Once you are satisfied with the fit, spread a little glue on two of the contacting miters and position one piece of side casing in place. Starting at the top, nail the casing to the jamb and wall studs *(right)*. Do not drive any nails near the bottom for now; you may need to adjust the casing slightly to fit the sill casing. Repeat for the other side casing.

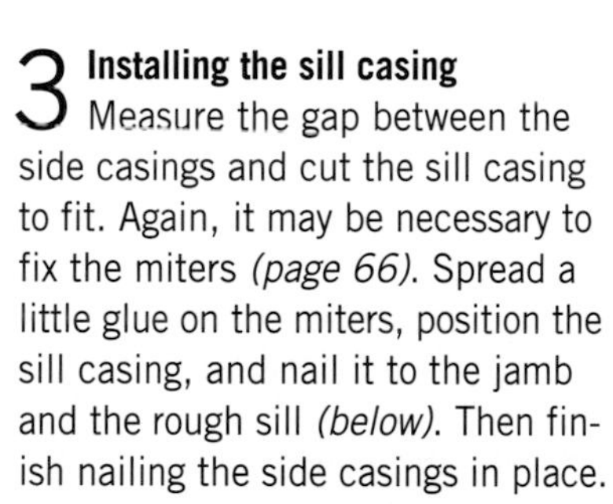

3 Installing the sill casing
Measure the gap between the side casings and cut the sill casing to fit. Again, it may be necessary to fix the miters *(page 66)*. Spread a little glue on the miters, position the sill casing, and nail it to the jamb and the rough sill *(below)*. Then finish nailing the side casings in place.

4 Cross-nailing the miters
To complete the installation, drive a nail into the edge of the side casing near the top so that the nail penetrates the head casing *(right)*. Repeat at the remaining three corners of the casing. This step will help ensure that the joints do not open with seasonal movement.

MODIFIED PICTURE-FRAME CASING

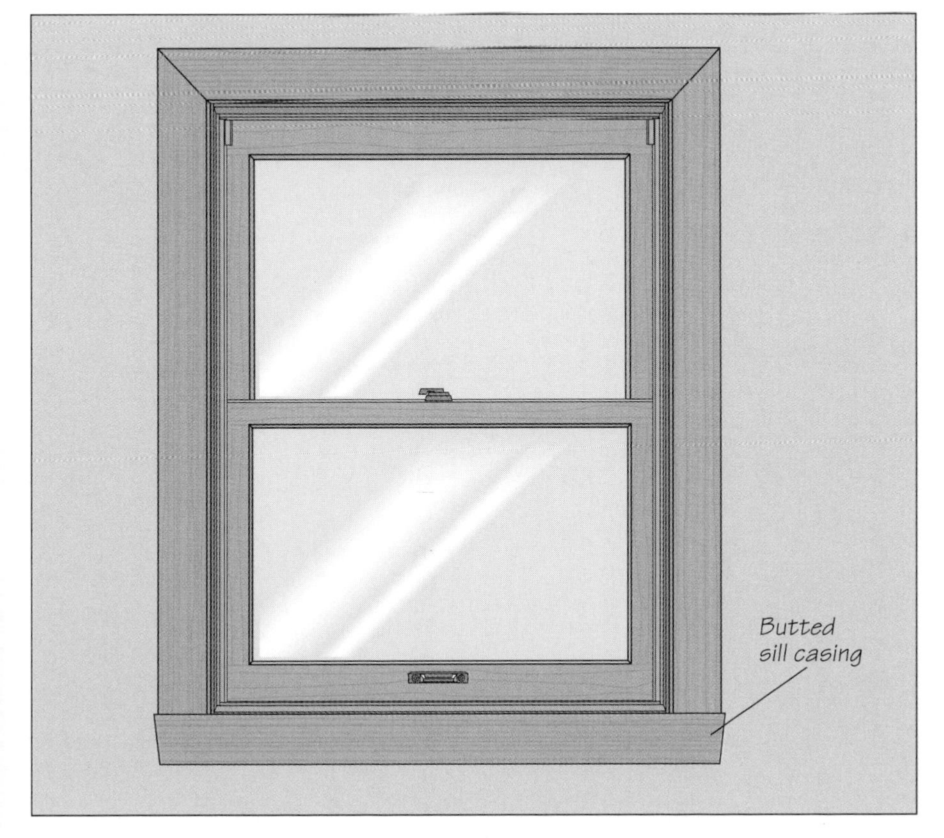

Installing a butted sill casing
You can simplify the installation of picture-frame casing by using butt joints at the bottom, instead of miters. As shown in the illustration at left, modified picture-frame casing involves miter joints at the top, but you can cut the bottom ends of the side casing square. Then crosscut both ends of the sill casing to span a little beyond the side pieces. For decorative effect, you can cut a shallow miter at each end of the sill casing. Then simply butt its thicker edge up against the side casings and nail it in place.

CORRECTING POOR-FITTING MITERS

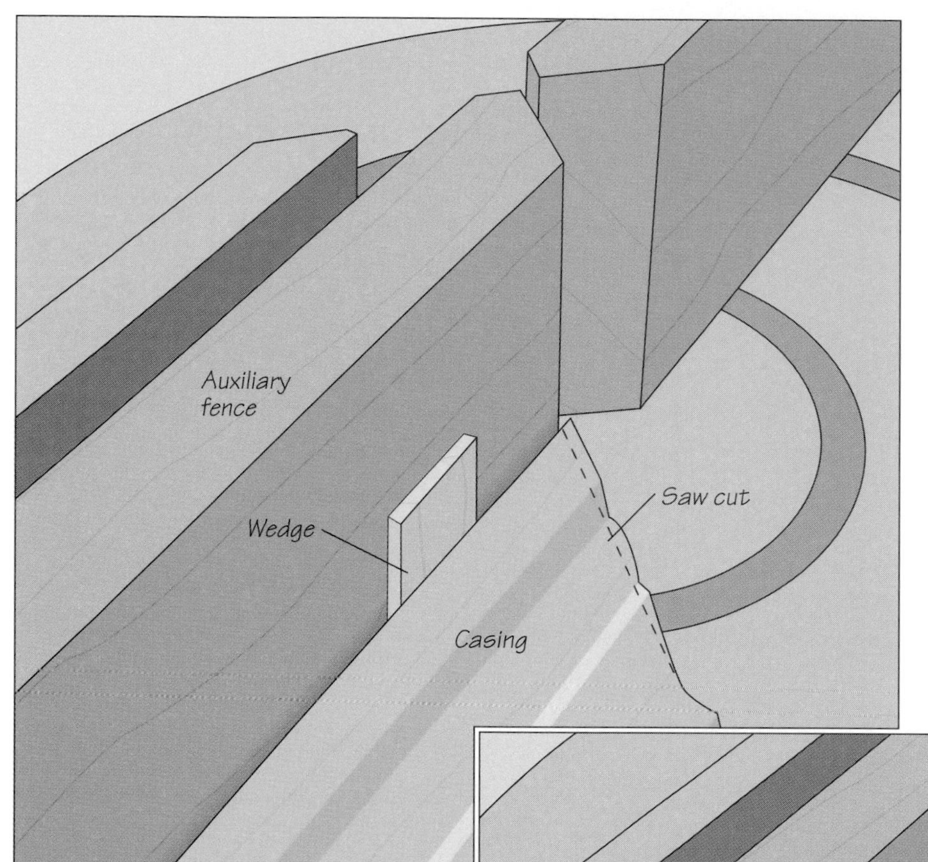

Closing a gap at the heel
You can fix a gap in a miter joint by adjusting the cutting angle on the power miter saw. But instead of resetting the saw's angle, it is simpler to change the angle of the workpiece on the fence. To close a gap at the heel of the miter, set the casing against the fence with the toe of the miter extending slightly beyond the fence. (In this case, an auxiliary fence has been attached to the regular fence to help line up the cut.) Then slip a thin wedge between the casing and the fence 1 or 2 inches from the end of the board. Now make the cut (represented by the dotted line in the illustration at left). Test-fit the joint and repeat the cut, if necessary, moving the wedge ¼ inch farther away from the end of the casing.

Saw cut
Wedge

Closing a gap at the toe
To close a gap at the toe of a miter, place the wedge 5 or 6 inches from the end of the casing and make the cut. As shown at right, the saw will shorten the heel of the miter. Test-fit the joint and repeat, if necessary, moving the wedge ¼ inch closer to the board end.

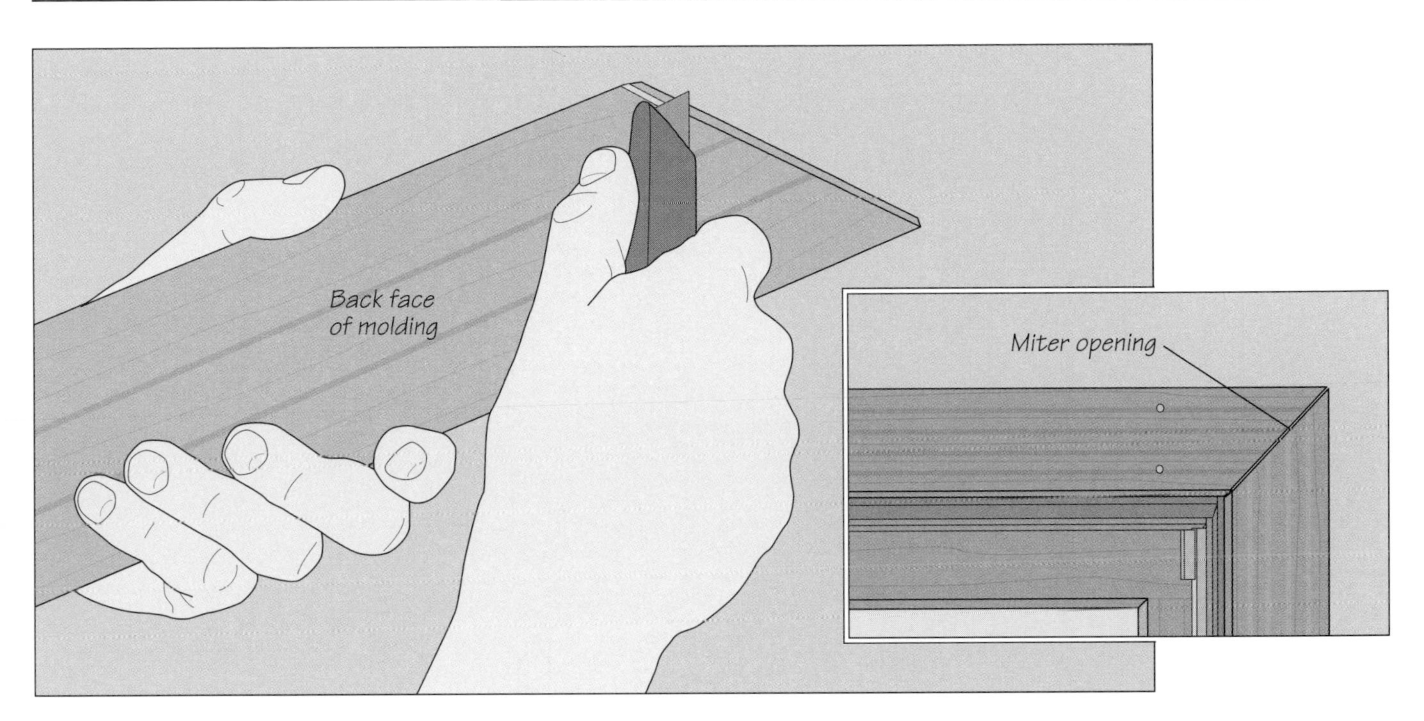

Back-cutting a miter

If a miter joint is open along its entire length *(inset)*, try the solutions described on page 66. If the joint is still open, remove some stock from the the back edge of one of the pieces. You can do this by repeating the corrective cuts on the power miter saw with a second shim placed under the casing to raise it slightly above the saw table, or by back-cutting the miter with a sharp utility knife, always cutting away from your body *(above)*. You can also use a block plane, as shown below.

Back-planing a miter

To back-cut a miter using a block plane, secure the casing in a vise so the miter is roughly parallel to the work surface. Holding the plane at an angle to the back edge of the miter, make a series of light cuts *(above)*.

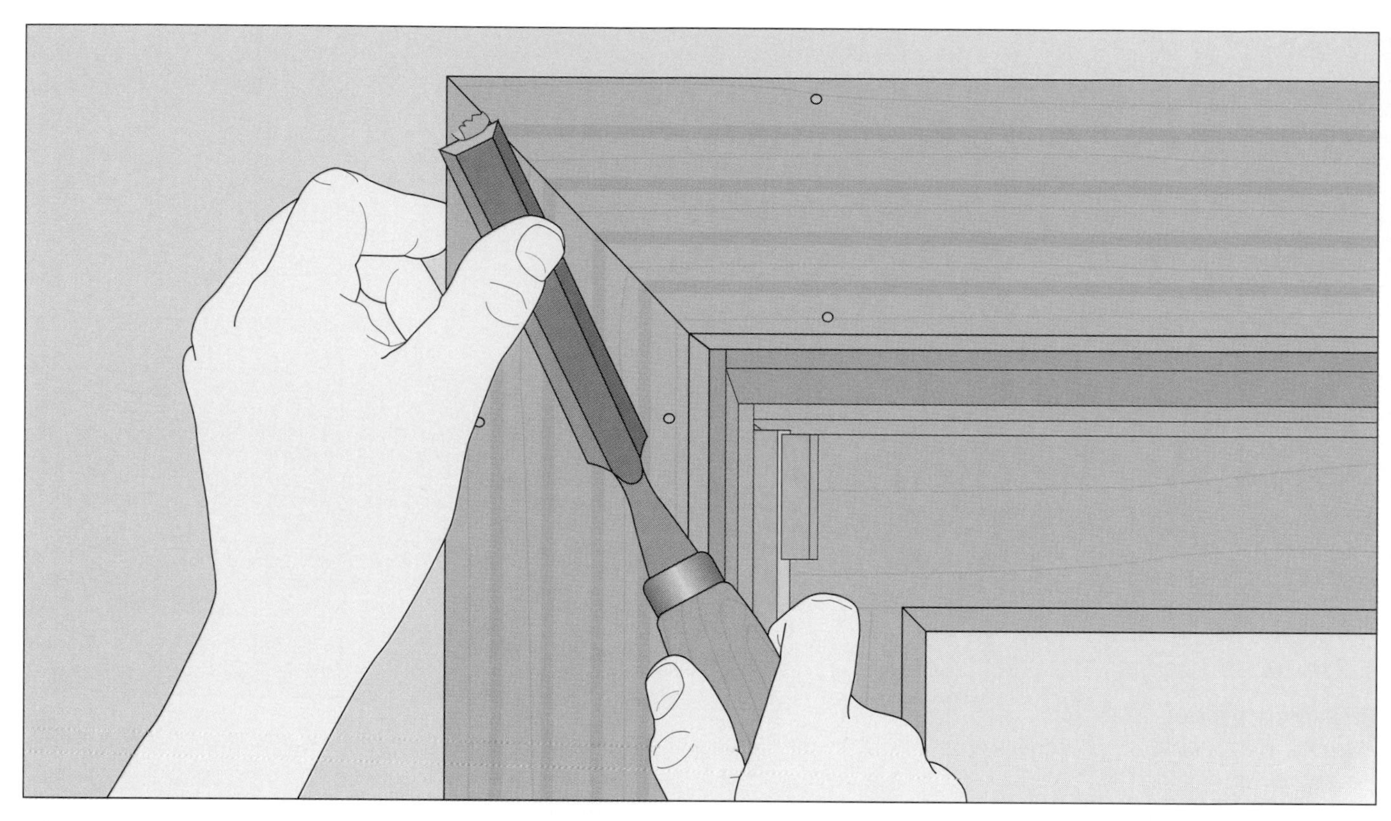

Trimming proud casing
A miter joint may fit well, but one of the mating pieces may be proud, or raised slightly above the level of the other. To remedy the problem, gently pare down the proud piece with a chisel *(above)*. Avoid sanding, which will leave a poor surface for finishing or painting, and which is much more difficult to do on molded casing. To avoid damaging the profile of intricate molding, you can install a shim behind the piece that is recessed as an alternative to chiseling the proud piece.

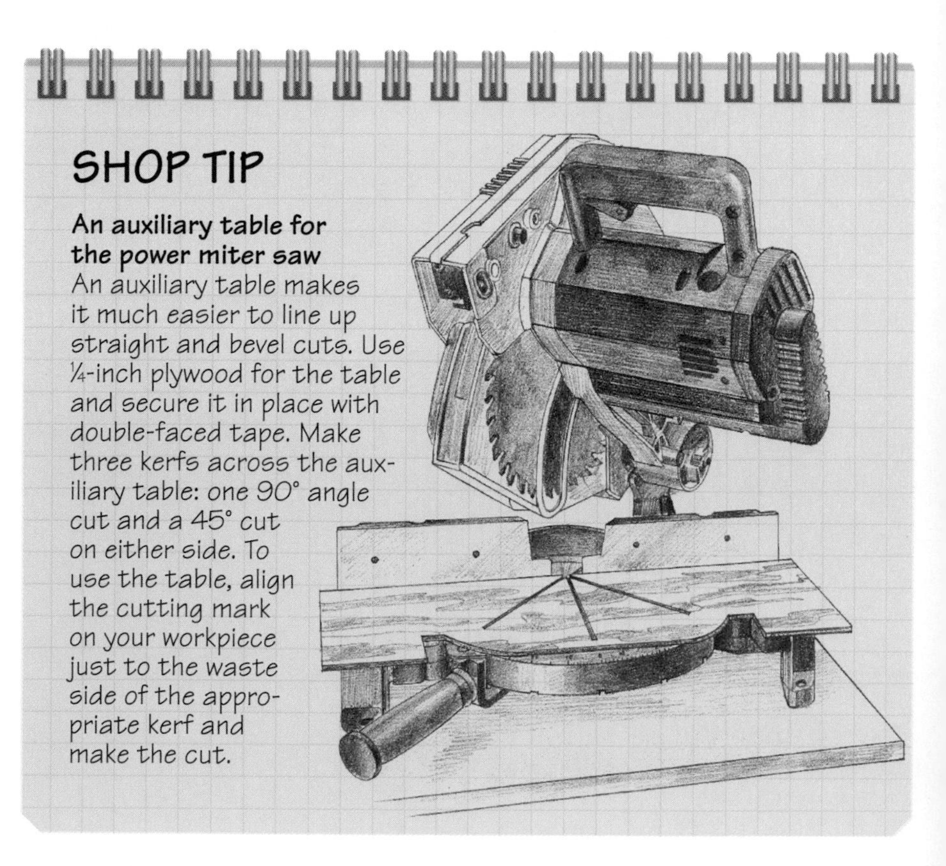

SHOP TIP

An auxiliary table for the power miter saw
An auxiliary table makes it much easier to line up straight and bevel cuts. Use ¼-inch plywood for the table and secure it in place with double-faced tape. Make three kerfs across the auxiliary table: one 90° angle cut and a 45° cut on either side. To use the table, align the cutting mark on your workpiece just to the waste side of the appropriate kerf and make the cut.

STOOL-AND-APRON CASING

Also known as "traditional" window casing, stool-and-apron casing is more difficult to make and install than picture-frame casing. However, its use of the butt joint allows different moldings to be combined for contrasting effect. The stool is cut to fit the window opening with two "horns" that extend beyond the side casing, typically by the same amount that the stool protrudes from the face of the casing. This can be anywhere between ¼ to ¾ inch, depending on the profile of the molding you are using. Lumberyards sell stool caps for assembly-line window installation, but you can easily make your own stool using a router. To balance the window, the head casing also extends past the side casings. The head casing can also include decorative rosettes *(page 73)*.

A window sill, or stool, is fastened to the wall studs with a finish nailer. Cut to fit the window opening, the stool features horns that extend past the window frame.

INSTALLING STOOL-AND-APRON CASING

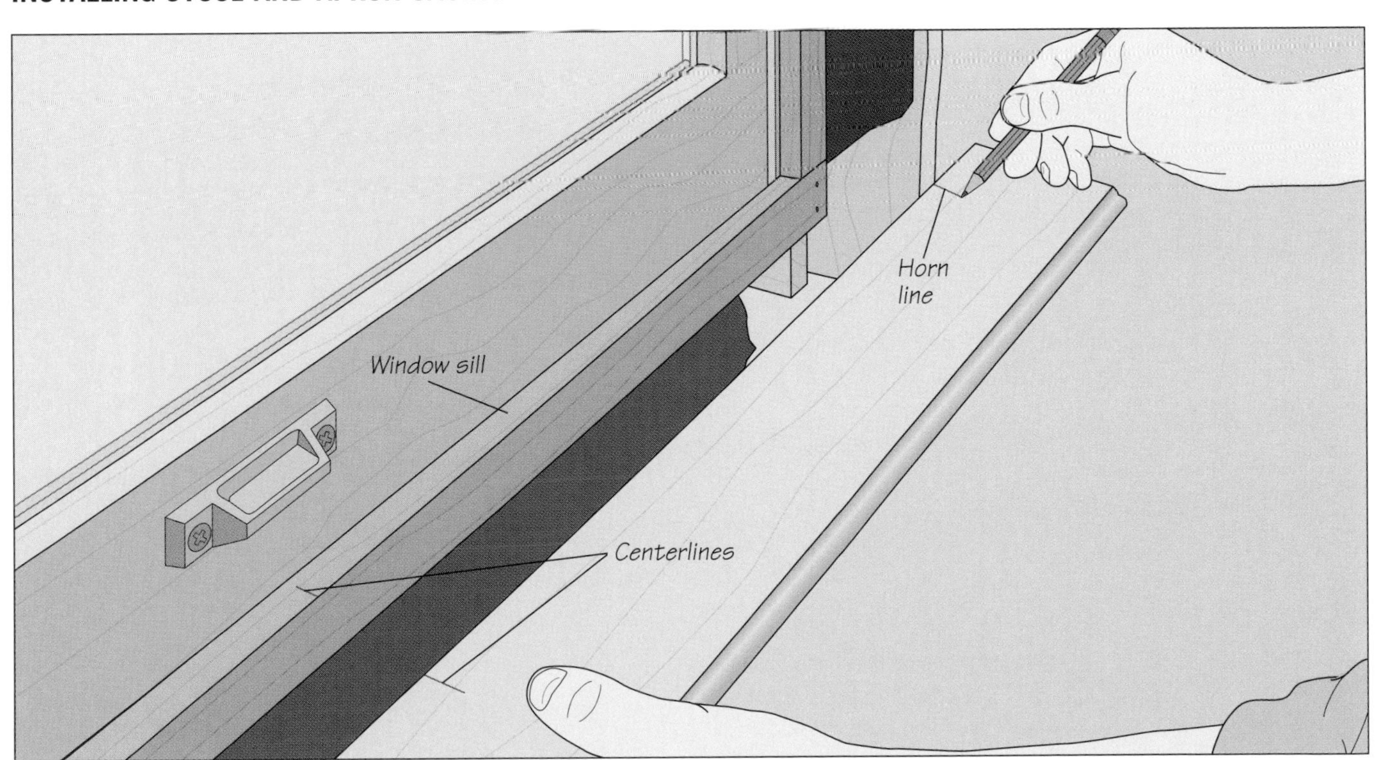

1 Marking the horns

Shape the outside edge of the stool and cut it to length. Mark the center of both the stool and the rough window sill. Then, mark the points on each side of the window where the wall meets the stool *(above)*. These will be your lines for cutting the inside edges of the horns. Extend both horn lines to the front edge of the stool with a try square.

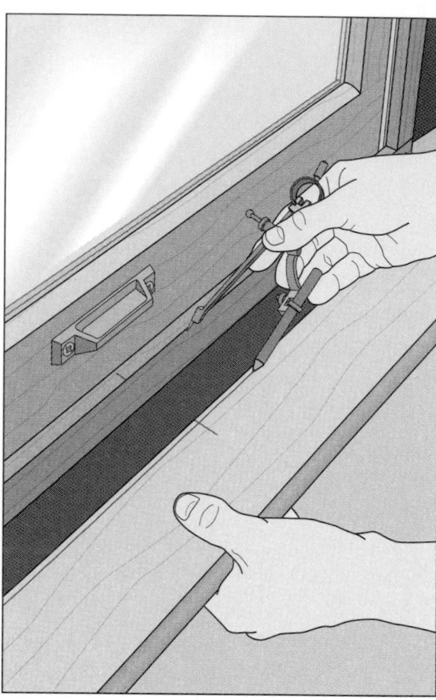

2 Marking and installing the stool
To finish marking your cutting lines for the horns, adjust a compass to the widest gap between the front edge of the rough sill and the drywall. Holding the stool against the wall with the center marks aligned, set the compass point at the edge of the wall and scribe a line for each horn *(above, left)*. To determine how much stock you need to trim from the inside edge of the stool, keep the same compass setting and mark a line along the length of the stool, running the compass point along the front edge of the rough sill *(above, right)*. Cut out the horns as well as the waste strip from the inside edge of the stool. Then nail the stool to the studs. If the window jambs are flush with the wall, install the side casing *(step 4)*. Otherwise, mount a modified jamb extension *(step 3)*.

3 Installing a modified jamb extension
Build a modified jamb extension *(page 61)* with no bottom piece. Fit the extension over the window jambs *(right)* and shim it in place, making sure that it is centered in the opening, square, and level. Then nail the extension to the jambs.

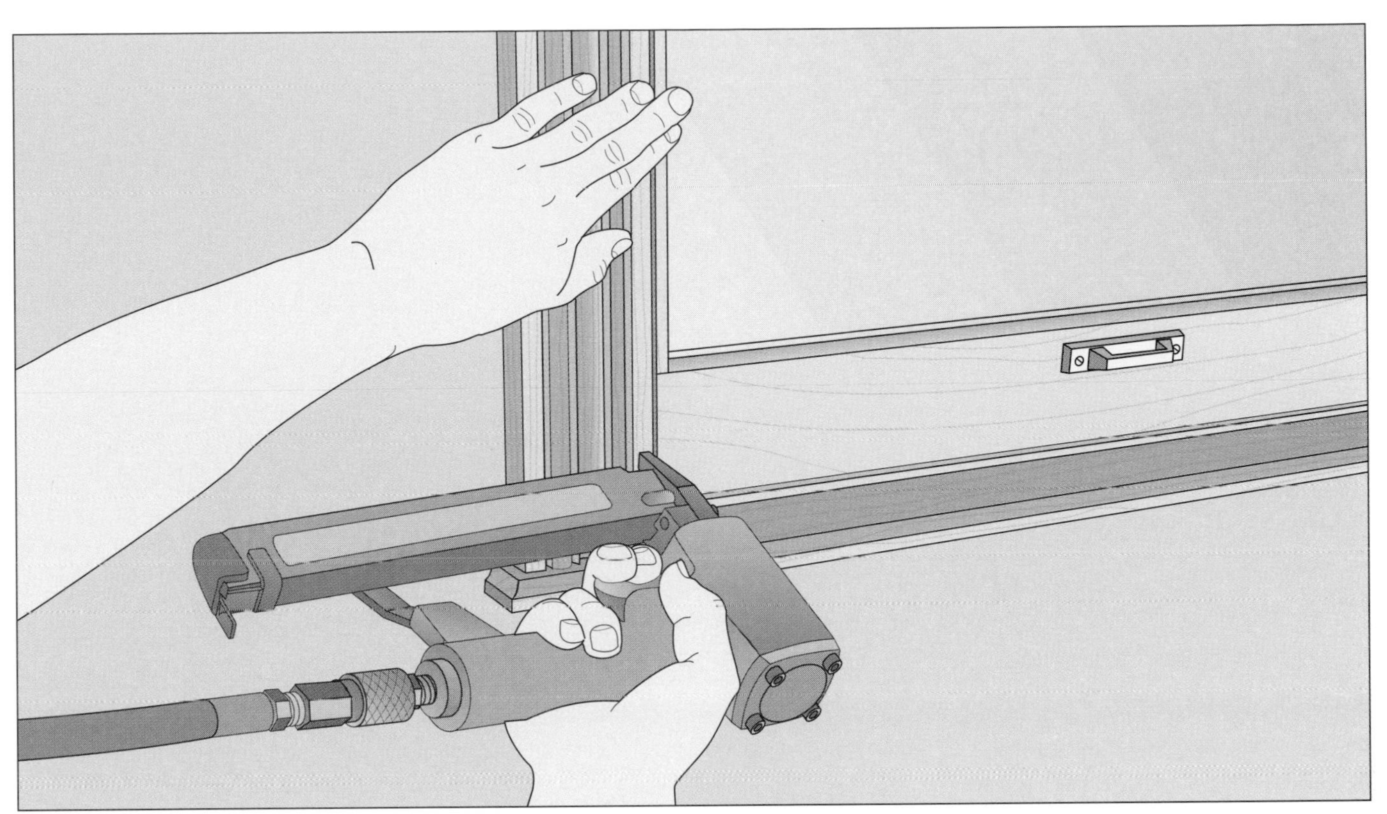

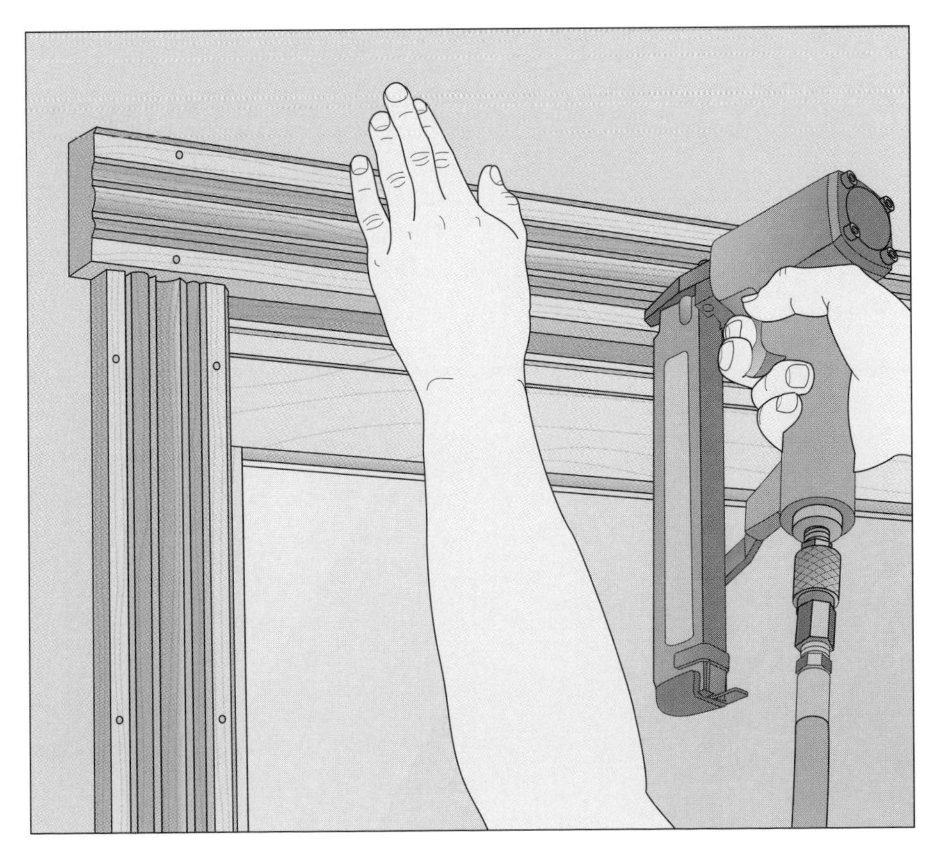

4 Installing the side casing
Mark the reveal around the jamb extension *(page 62)* to match the reveal between the stool and rough sill. Then cut the side casings to length, sawing both ends square. Install the side casings as you would for picture-frame casing *(page 65)*, nailing them into the window jamb and studs *(above)*. Space the nails 6 to 8 inches apart.

5 Installing the head casing
Cut the head casing to the same length as the stool, center it on the side casing pieces, and nail it in place *(left)*. Drive the nails into both the head jamb and the rough header every 6 to 8 inches.

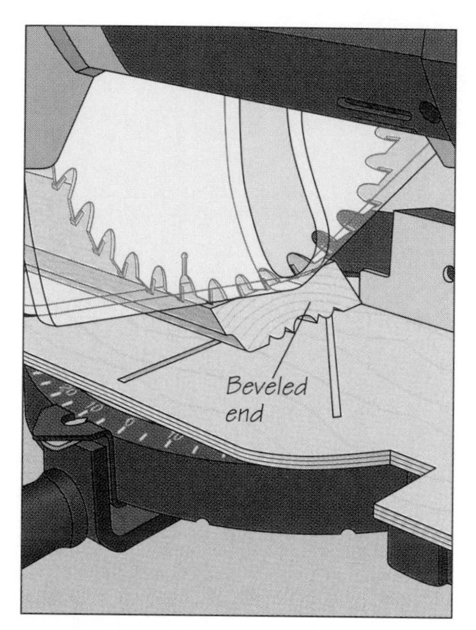

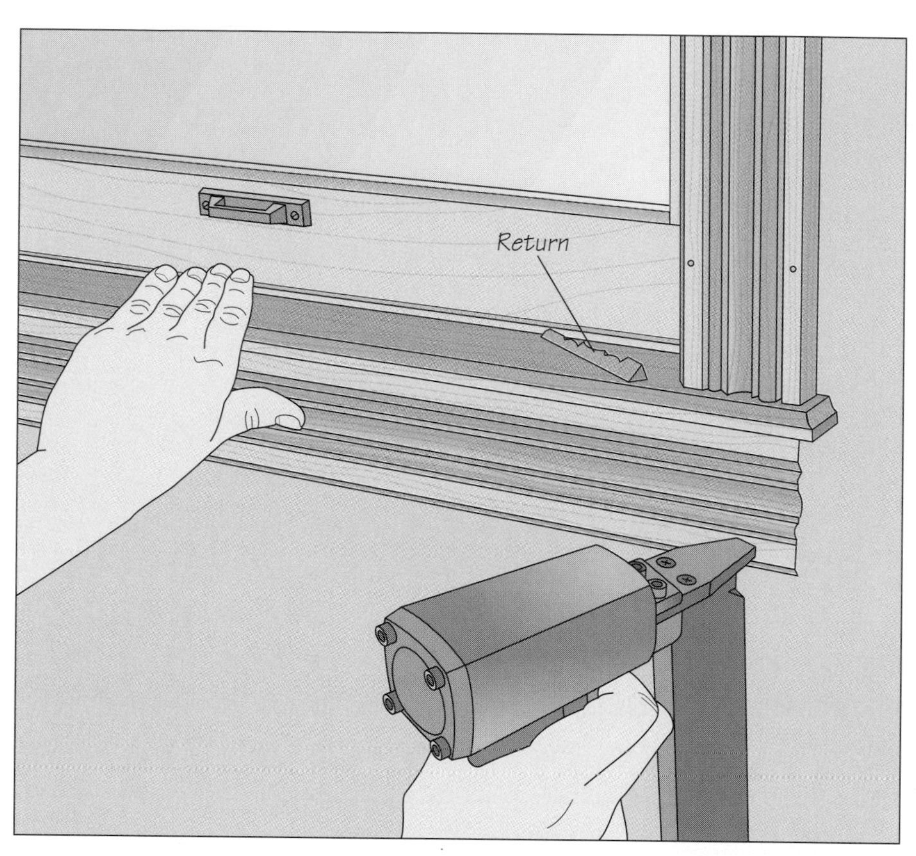

6 Installing the apron and returns
Once the stool and casing are in place, complete the window by installing the apron beneath the stool. Start by measuring the distance between the outside edges of the side casings and cut the apron to your measured length, sawing 45° bevels at both ends of the apron. To conceal the end grain, glue on matching end pieces, known as returns. Make them on the power miter saw by cutting a 45° bevel in a piece of scrap molding with the same profile as the apron, then cutting off a narrow wedge of stock at the end of the piece *(above, left)*. Nail the apron into the rough sill and the wall studs *(above, right)*, then glue the returns to the ends of the apron *(page 57)*.

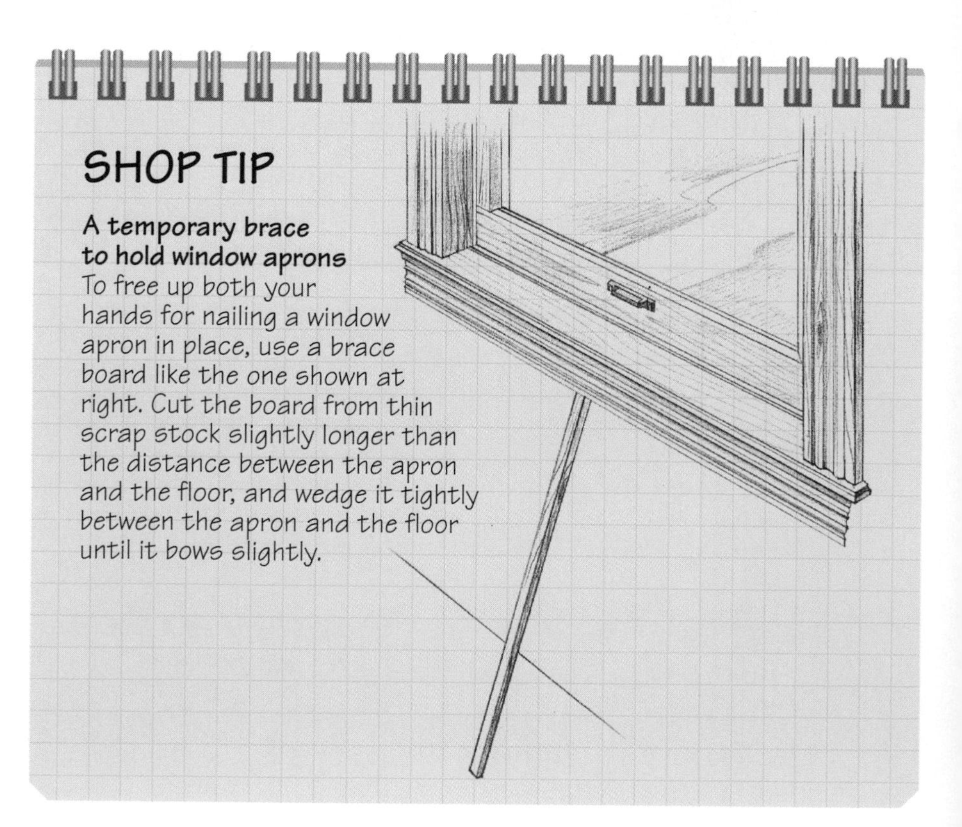

SHOP TIP

A temporary brace to hold window aprons
To free up both your hands for nailing a window apron in place, use a brace board like the one shown at right. Cut the board from thin scrap stock slightly longer than the distance between the apron and the floor, and wedge it tightly between the apron and the floor until it bows slightly.

MAKING AND INSTALLING ROSETTES

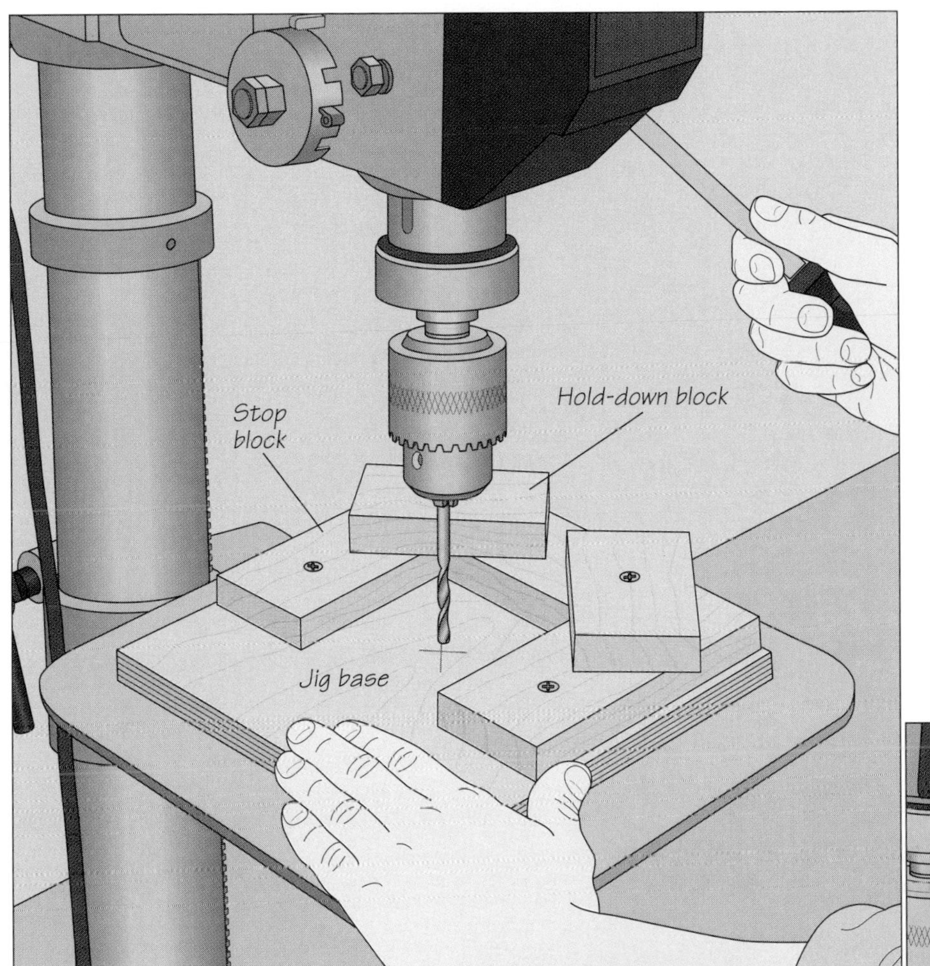

1 Making a drilling jig
Size blanks for the rosettes *(page 58)* so that they are slightly wider than the casing, and their outside edges will align with the ends of the stool. You can also make the blanks thicker than the casing. To cut the rosettes on the drill press, make a drilling jig to secure the blanks. Center a blank on a ½-inch plywood base and butt stop blocks of the same thickness around three sides of the blank and screw them to the base. Screw down two more blocks as hold-downs at a angle to the corners formed by the stop blocks. Mark the center of the jig and set it on the drill press table. Install a brad-point bit, align the jig's centerpoint directly under the bit *(left)*, and clamp the jig to the table.

2 Securing the blank to the jig
Install a rosette cutter in the drill press and set the machine's drilling speed following the manufacturer's instructions. Place a blank in the drilling jig and lock it in place by clamping a notched hold-down block over it *(right)*.

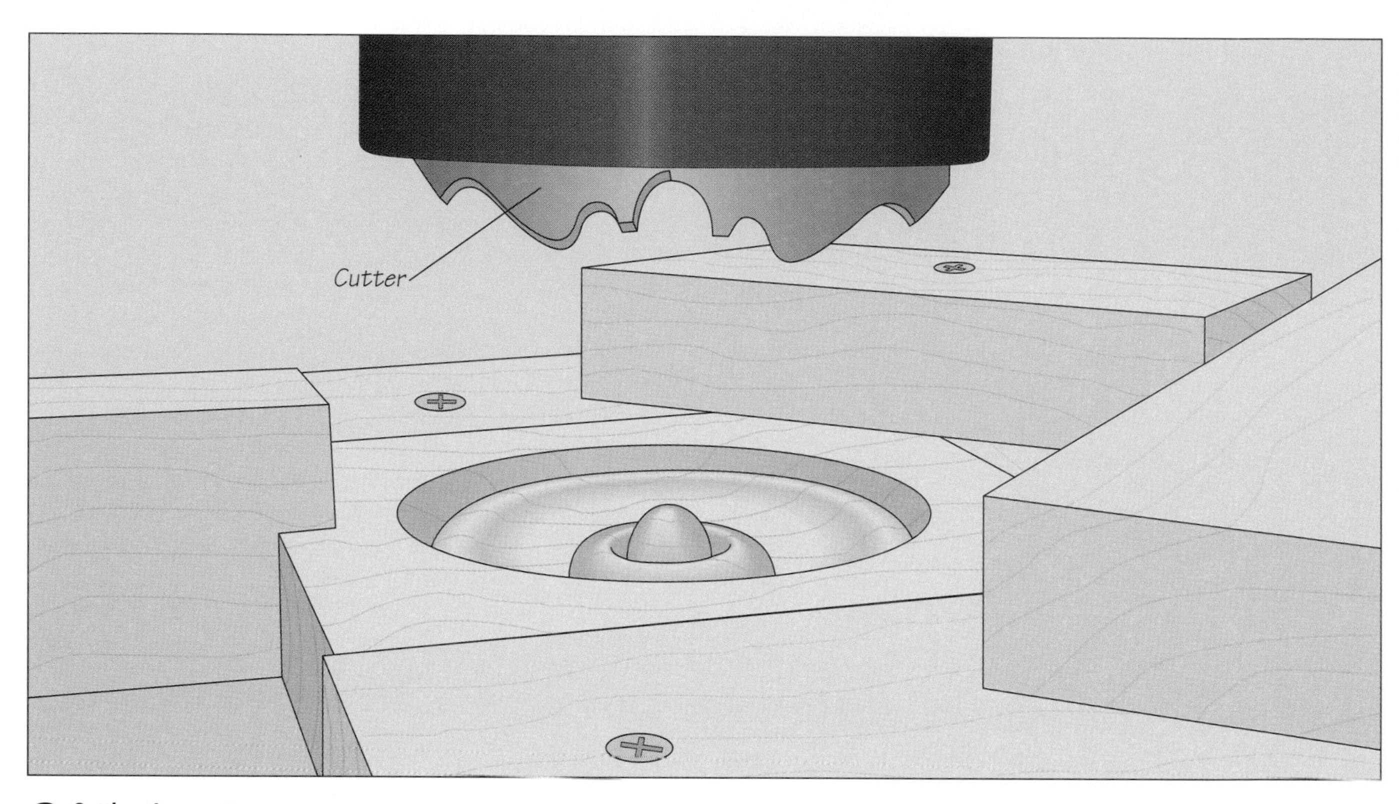

3 Cutting the rosettes
Turn on the drill and lower the quill until the cutter lightly contacts the wood. Continue cutting until the rosette has the desired profile *(above)*. Install the rosettes with the end grain on the top and bottom after nailing the side casings in place *(page 71)*; then cut the head casing to fit between the rosettes.

SHOP TIP

A shop-made rosette cutter
You can cut rosettes by modifying a drill press fly cutter with a beading blade from a table saw's molding head. Notch the fly cutter arm to accommodate the beading blade, locating the cutter about 1 inch from the end of the arm. Make sure it fits securely in the notch so it cannot shift during use. Bore a hole through the arm and use a bolt, washer, and nut to fasten the blade in place, its flat face toward the direction of cutter rotation. Use a similar drilling jig to the one shown on page 73 to hold your rosette blanks, and make the cuts as you would with a commercial rosette cutter.

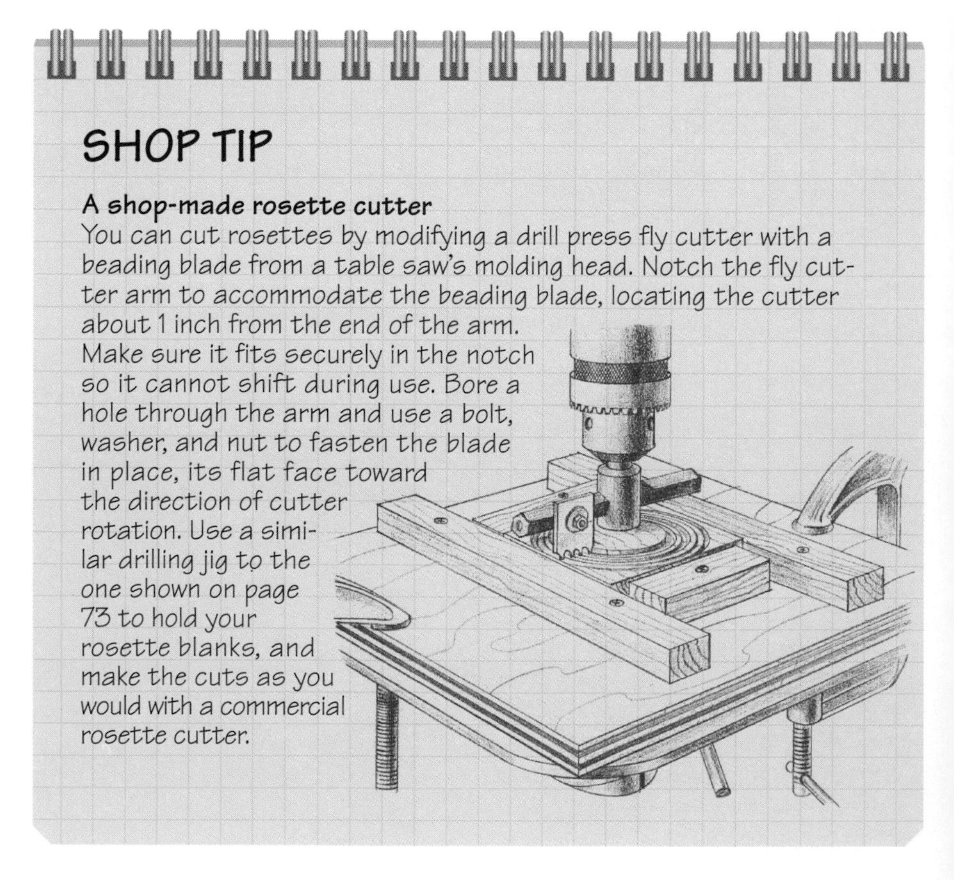

MAKING A WINDOW SASH

ASSEMBLING THE SASH

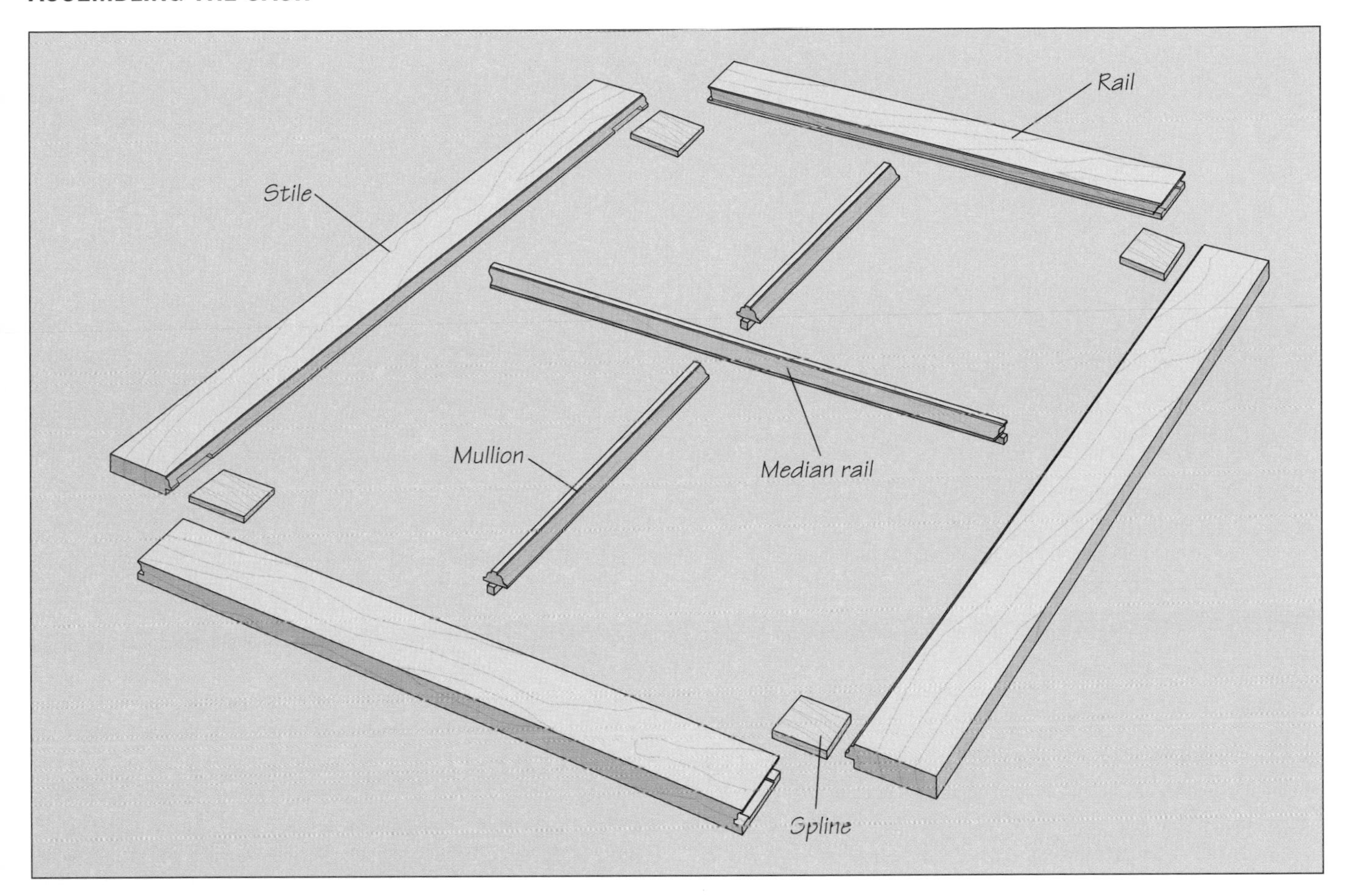

Paired with a shop-made mortising jig, a router cuts a mortise in one of the stiles of a window sash. The jig ensures that the mortise is centered on the edge of the stock. A matching mortise will be cut in the end of the adjoining rail and a spline will reinforce the joint between the two pieces.

1 Planning the job
The window sash shown above consists of two vertical stiles, two horizontal rails, a median rail, and two mullions that divide the sash vertically. The pieces are connected by cope-and-stick joints cut on a shaper. The joints between the stiles and rails are reinforced by splines. To size your stock, make the stiles equal to the height of the opening for the sash. For the rail length, take the width of the opening and subtract twice the stile width. Then add twice the depth of the coping cuts you will make *(step 2)*. If, for example, the width of the window opening is 32 inches, the stiles are 3¼ inches wide, and the depth of the coping cuts is ¼ inch, each rail should be 26 inches long. You can also make the bottom rail wider than the other pieces to accommodate handles. To determine the length of each mullion, take the height of the opening and subtract the width of the three rails. Then add four times the depth of the coping cuts. Divide the total by two.

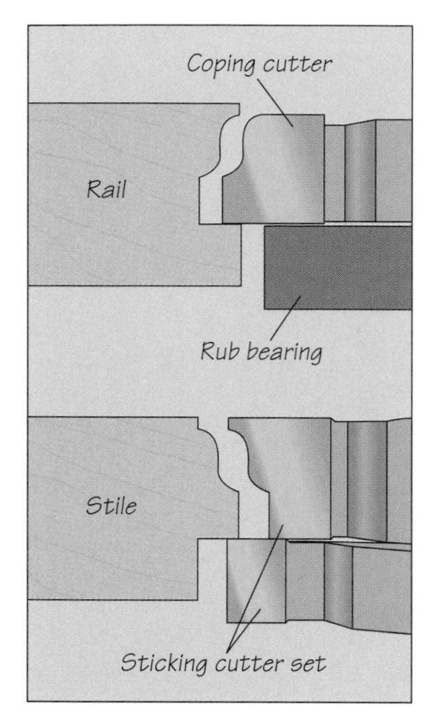

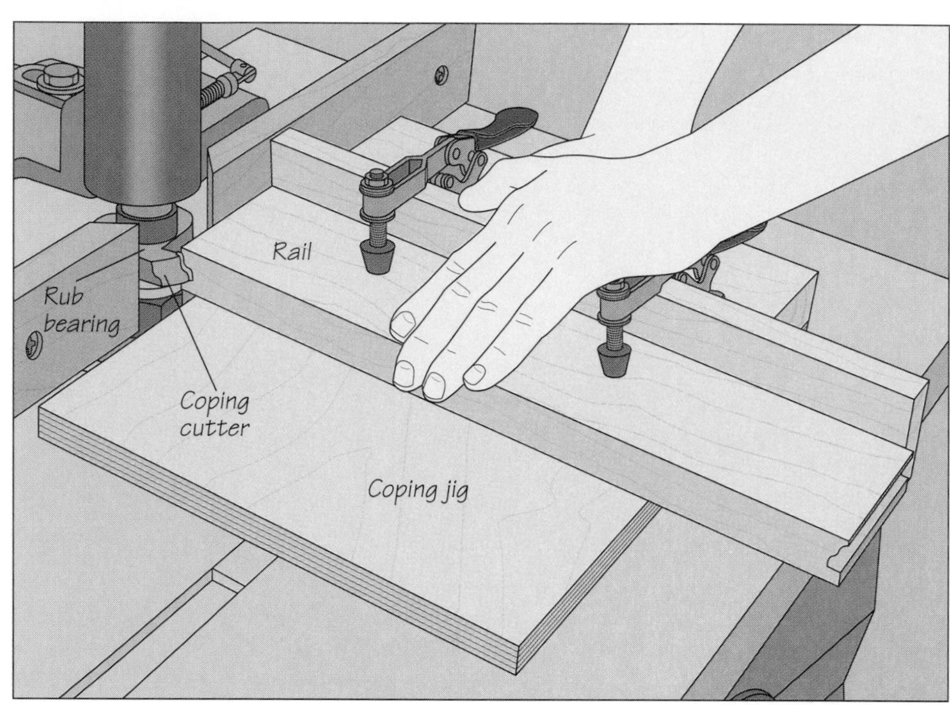

2 Making the coping cuts
As shown above at left, the joinery for the rails and stiles is done by matching cutters on the shaper. The coping cutter is used on the ends of all rails and mullions. The sticking cutter shapes the inside edges of the sash pieces. To set the height of the coping cutter, first install the sticking cutter in the shaper and adjust its height so the top of the cutter is level with one of the sash pieces set face down on the table. Make a cut *(step 5)* in a test piece the same thickness as the sash stock. Then install the coping cutter and rub bearing on the shaper and butt the cut end of the test piece against the cutter to set its height. For the coping cuts, position the fence slightly behind the rub bearing and build a coping jig *(page 90)*. Use the jig to feed both ends of the rails into the cutter *(above, right)*. To cope the end of the median rail and mullions, shape a wide piece and rip the widths you need on the table saw *(step 3)*.

3 Ripping the median rail and mullions
Once you have made the coping cuts on two wide boards for the median rail and mullions, position the table saw rip fence for cutting the median rail—typically one-third the width of the stiles. Feed the board into the blade with a push stick *(right)*. Reposition the fence for the mullions and cut them from the other board the same way.

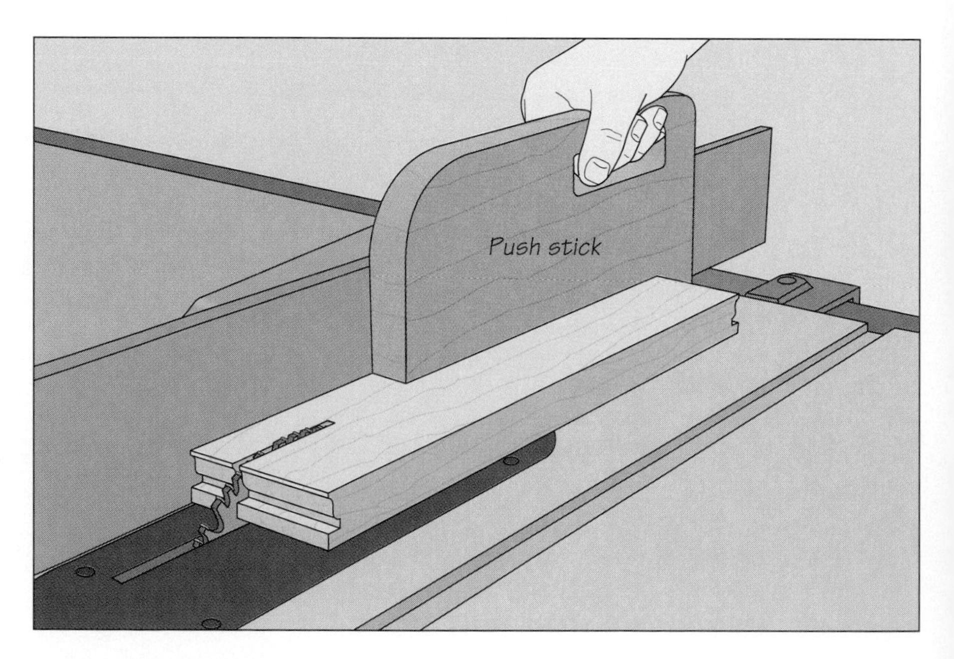

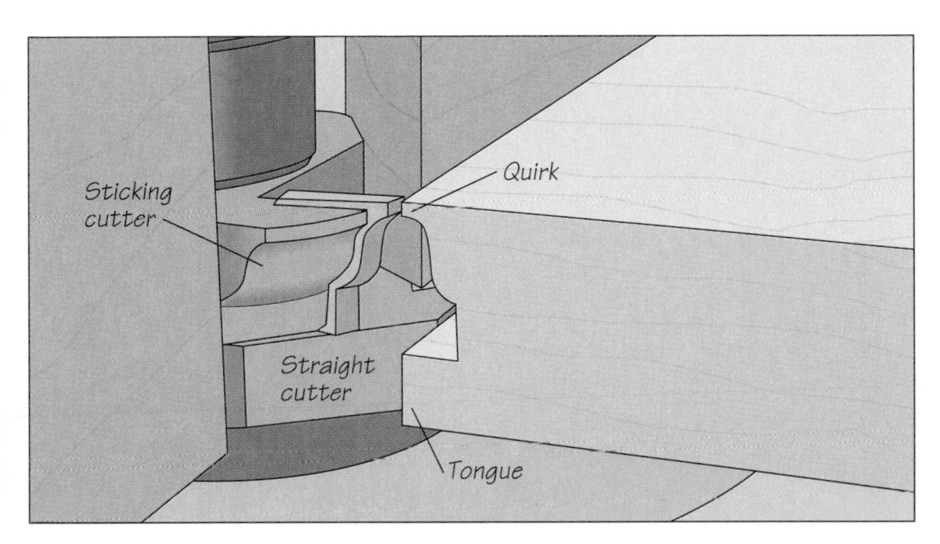

4 Setting up the sticking cutter
Once all the rails and mullions are prepared, remove the coping cutter and rub bearing from the shaper and install a sticking cutter set. The one shown features a straight cutter, which should be the same width as the tongue left by the coping cuts. This setup will shape the inside edge of all the sash pieces and cut rabbets to support the glass. Butt one of the rails against the bit to set the height of the sticking cutter *(left)*; the tip at the top of the cutter should be aligned with the lip, or quirk, at the top of the coped end.

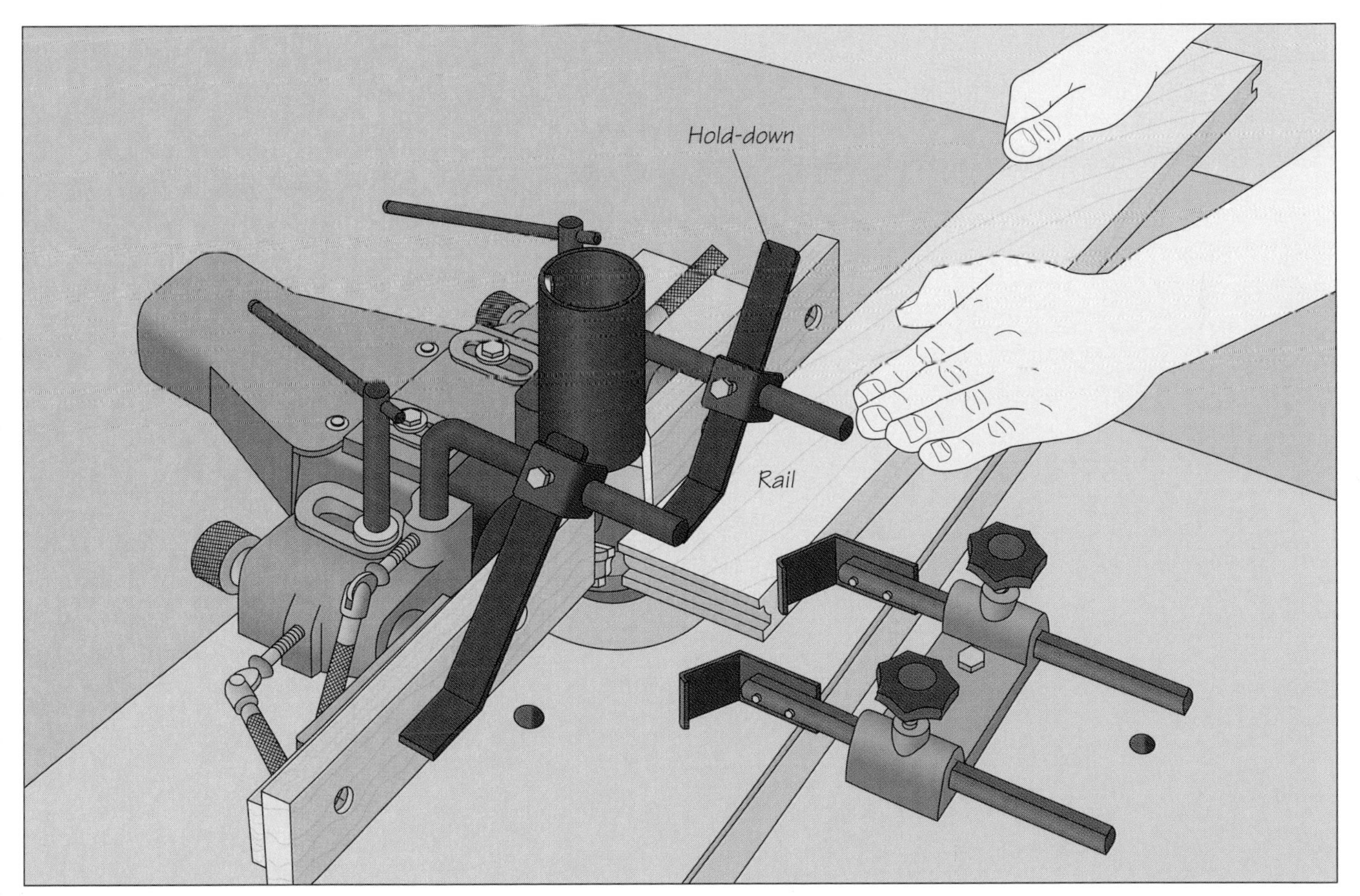

5 Making the sticking cuts
To make the sticking cuts, adjust the shaper's fence to make a full cut in the edge of the stock; the cutter should just touch the widest point of the workpiece. Also install commercial hold-downs on the shaper's fence and table to secure the stock during the cuts and prevent kickback and chatter. Then, make the sticking cuts in the inside edges of the rails and stiles, feeding the stock at a steady rate *(above)* and use a push stick to finish the cuts. Repeat this process for the median rail and mullions but this time shaping both edges of the piece.

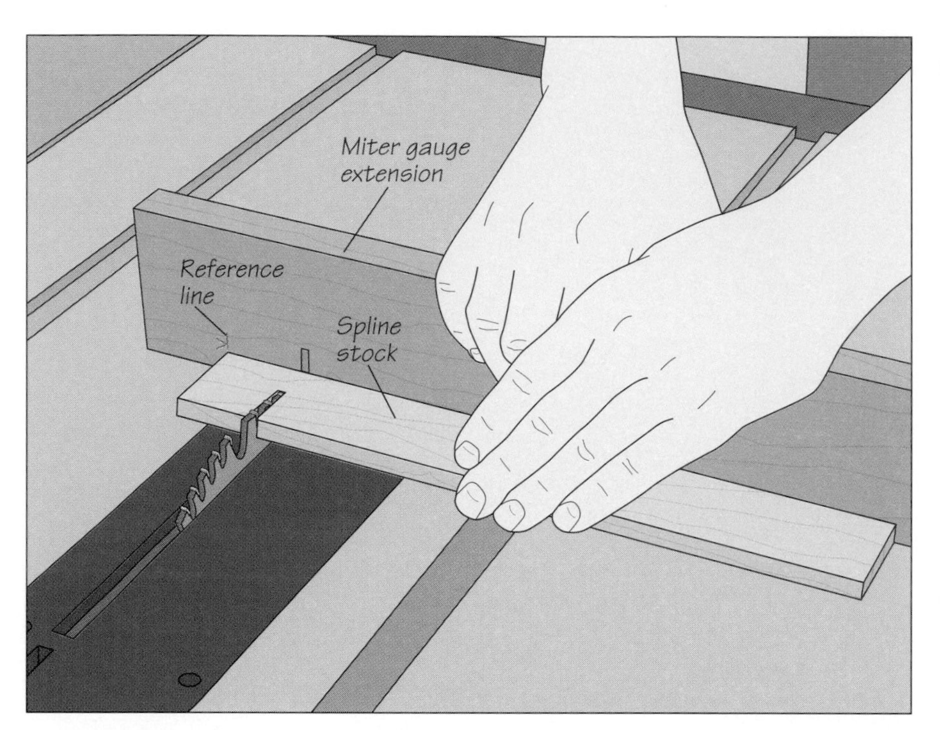

6 Strengthening the corner joints
Reinforce the joints between the stiles and the top and bottom rails with splines. Start by routing mortises for the splines in the ends of the rails and the inside edges of the stiles *(page 75)*. The spline should fit the mortises snugly and be shorter than the combined depth of the two mating mortises. You can cut all the splines from a single board. To do the job on the table saw, screw a wooden extension to the miter gauge. Ensure that all the splines will be the same length by marking a reference line on the extension. Align the end of the board with the line and hold its edge against the extension to cut each spline *(left)*. For maximum strength, cut the splines so their grain will run in the same direction as the grain of the rails.

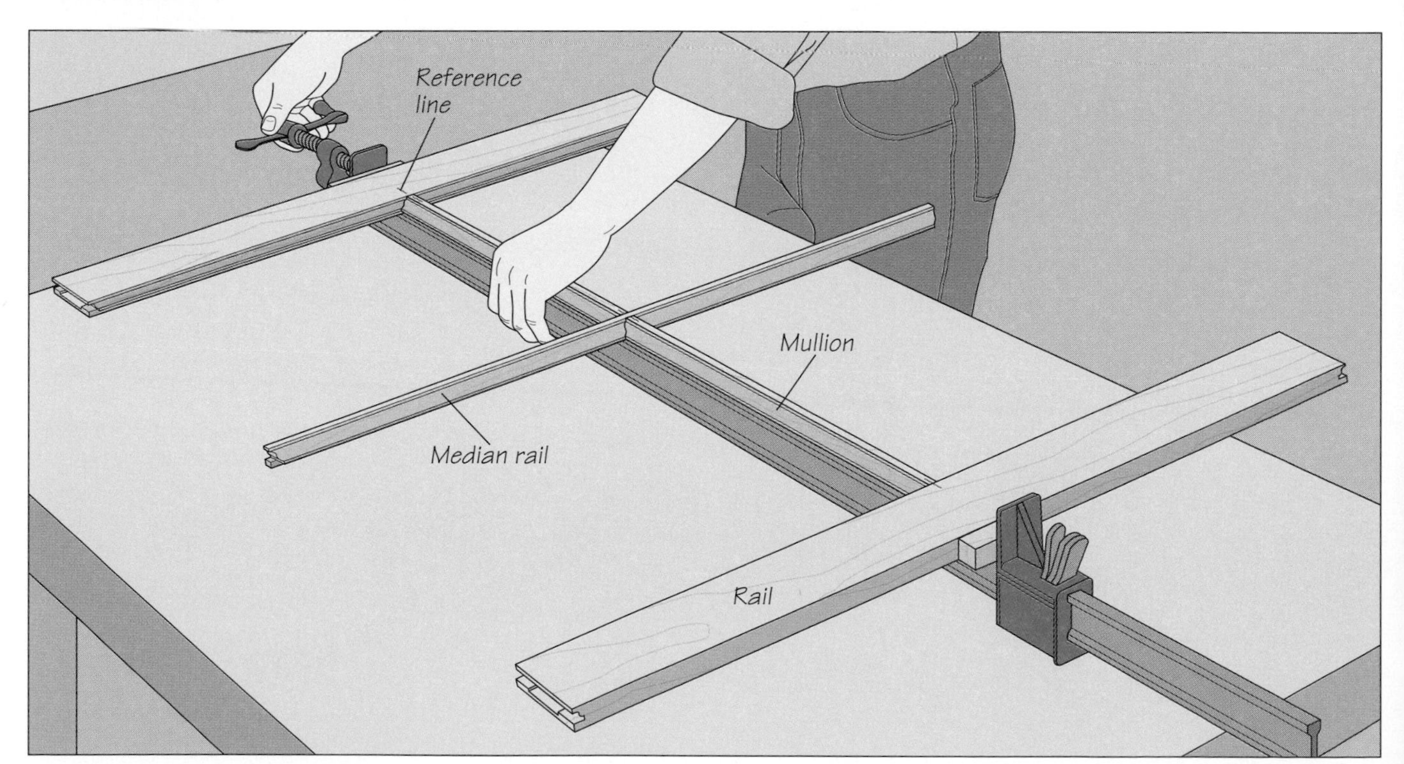

7 Gluing the mullions to the rails
Glue up the frame in two steps. Start by gluing the rails and mullions together, as shown above, then add the stiles *(step 8)*. Test-fit the pieces, marking reference lines across the joints with the mullions to help you align the parts during glue-up. For the rails and mullions, apply glue to the contacting surfaces of the boards. Assemble the pieces and install a bar clamp to secure the mullions to the top and bottom rails *(above)*; use wood pads to protect the stock.

8 Gluing the stiles to the rails
Insert the splines in the rails and apply glue to the joints between the rails and stiles. Spread glue in the mortises and onto the splines. Turn the window over and secure the stiles in place with bar clamps *(above)*. Align a clamp with each rail, ensuring that the ends of the stiles are flush with the edges of the rails. Use wood pads to protect the stock. As soon as the clamps are tight, check the assembly for square by measuring the sash from corner to corner in both directions. The two diagonals should be equal. If not, readjust the clamping pressure slightly until the sash is square.

INSTALLING THE GLASS AND GLASS-STOP MOLDING

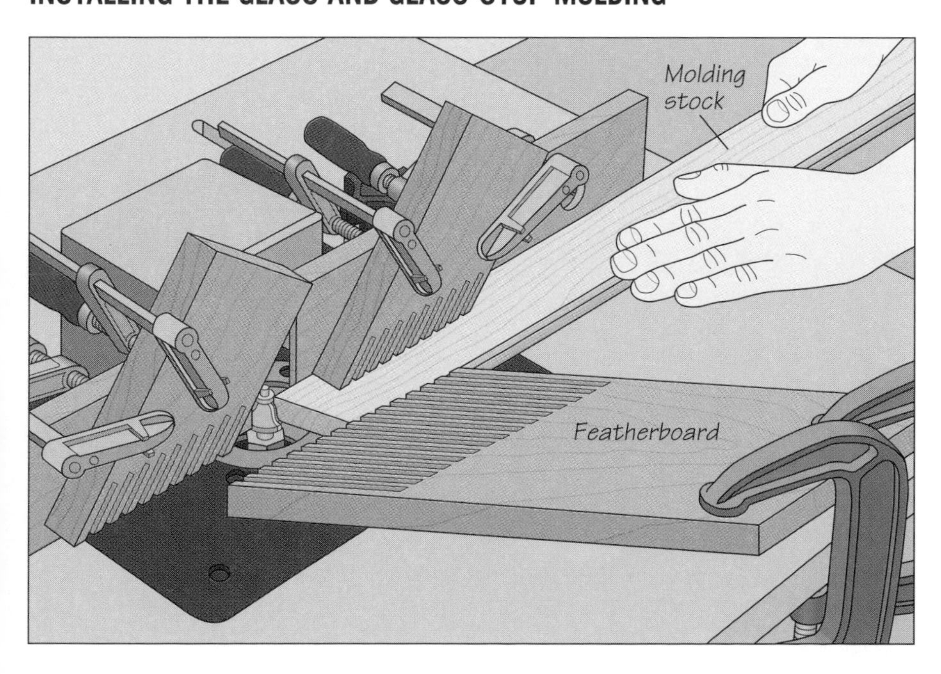

1 Making glass-stop molding
Glass-stop molding will hold the glass against the rabbets in the window sash. To prepare the molding with a router, install a decorative molding bit in the tool and mount it in a table. Shape both edges of a wide board long enough to yield all the molding you will need, then rip the molding strips from the stock. Use three featherboards to support the workpiece during the cut: two clamped to the fence on either side of the bit and one clamped to the table. Feed the board into the cutter while keeping it flush against the fence; finish the pass with a push stick. Repeat to shape the other edge *(left)*. Cut the molding off the board on the table saw, then saw it to length, making 45° miter cuts at the end of each piece.

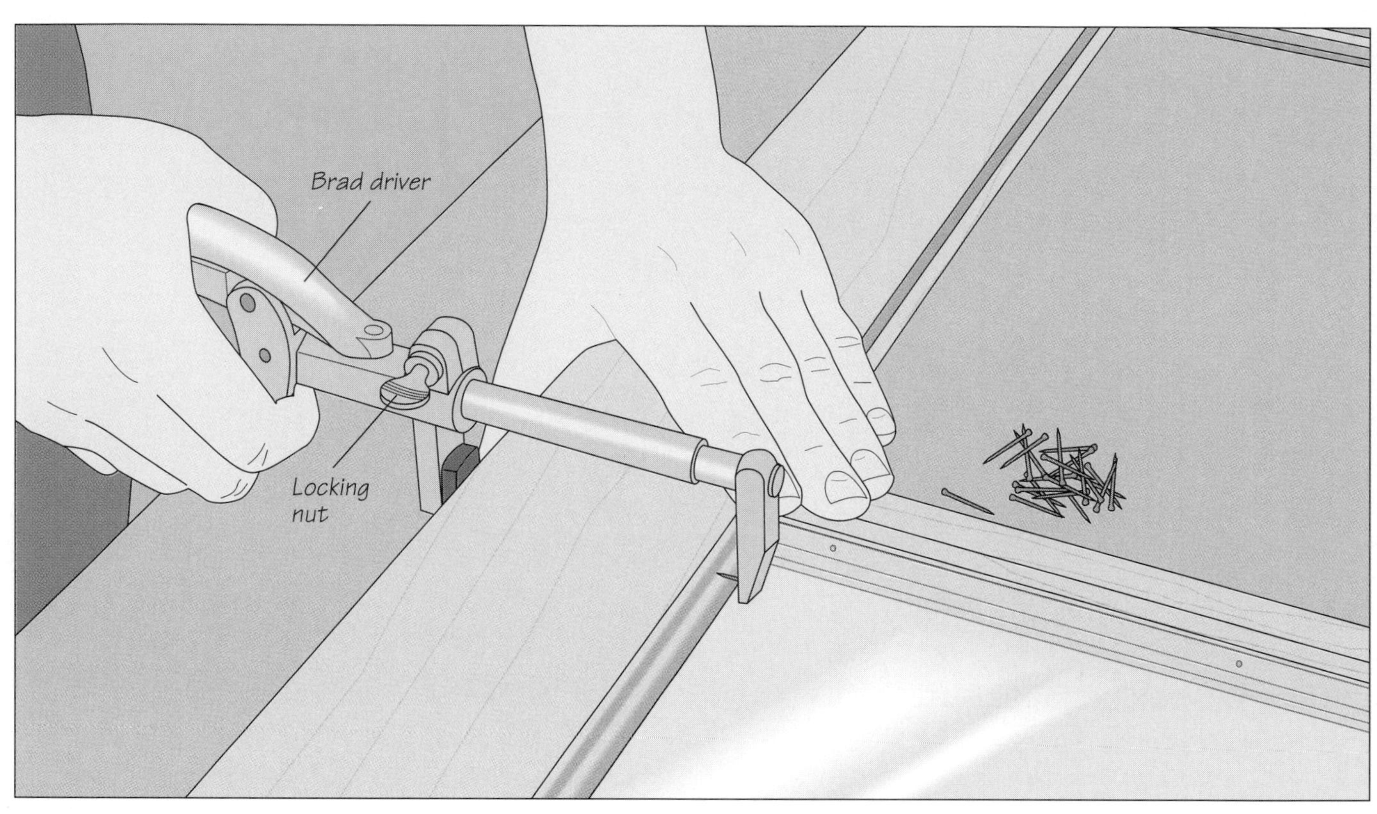

2 Installing the glass and the molding Set the sash and glass on a work surface, then place the molding in position. Bore a pilot hole every 6 inches, insert a finishing nail into each hole, and drive it home. To use a brad driver, as shown above, adjust the jaws against the sash and the nail, then tighten the locking nut. Holding the sash steady, squeeze the jaws to set the nail.

SHOP TIP

Installing the molding with a hammer
If you are using a hammer to nail glass-stop molding in place, protect the glass by placing a piece of cardboard on it as you drive each nail, as shown here.

A GLAZING BAR HALF-LAP JOINT

The glazing bar half-lap joint shown at right forms a stronger bond than the cope-and-stick joint for connecting the mullion and median rail of a divided window sash. The pieces, called glazing bars, are joined by mitered half-laps. Rabbets are cut along the back edges of the bars to accommodate the glass and glass-stop molding. The ends of the bars are joined to the rails and stiles with cope-and-stick joints.

MAKING A GLAZING BAR HALF-LAP JOINT

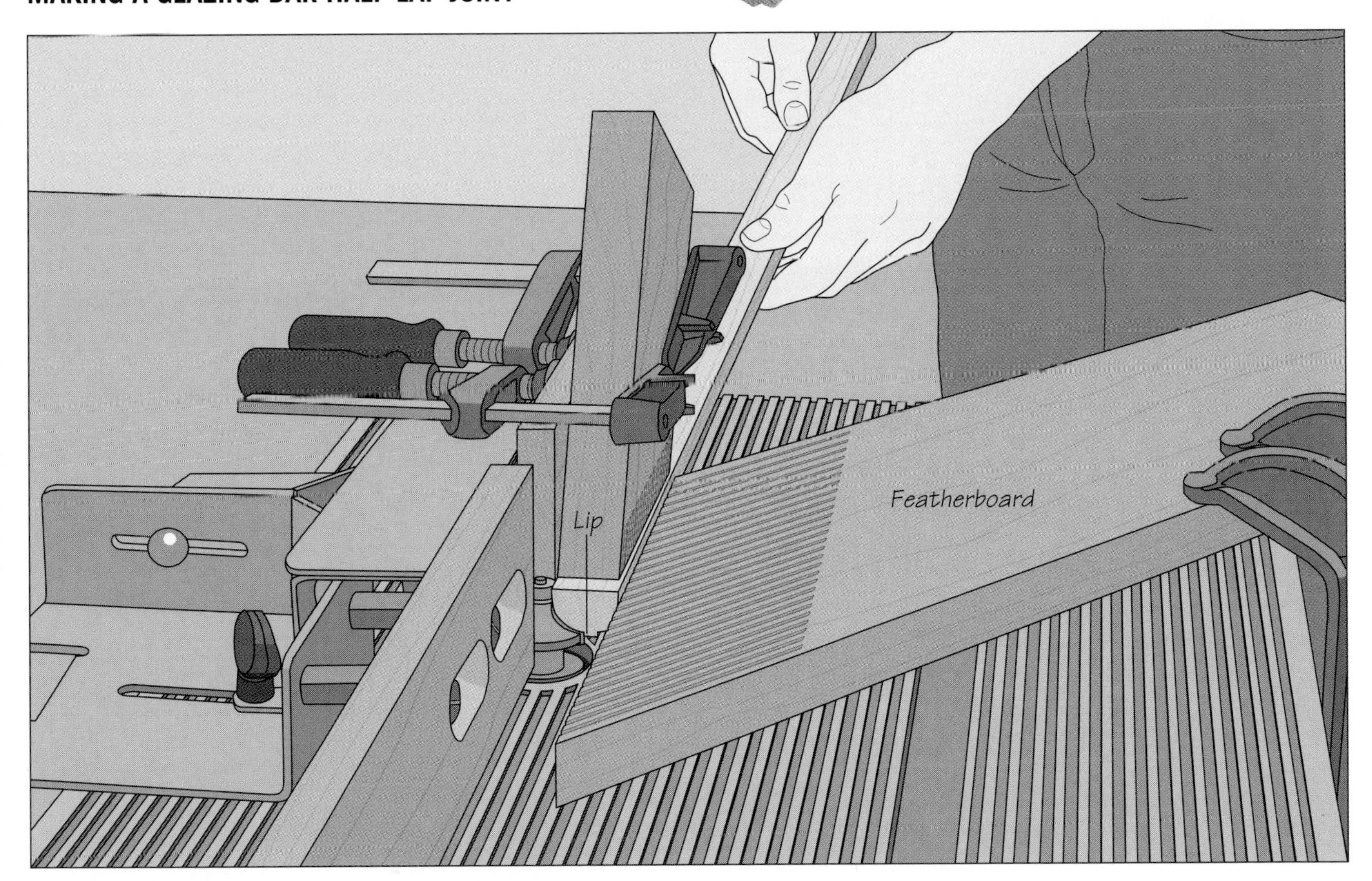

1 Molding the glazing bars
The joint is made in three stages: Start by cutting the proper profile into the glazing bars, as shown above; next, cut rabbets into the opposite side of the bars to hold the glass and molding strips *(step 2)*; finally, produce the mitered half-lap *(steps 3 to 5)*. For the first stage, install a piloted round-over bit in a router, mount the tool in a table, and align the fence with the bit's pilot bearing. The stock should be wide enough so that making a pass on each side of the bar will leave a ¼-inch-wide lip between the cuts. Support the workpiece during the operation with three featherboards: Clamp one to the table opposite the bit and two to the fence on each side of the cutter. (In the illustration, the featherboard on the outfeed side of the fence has been removed for clarity.) Feed the bar into the bit until your fingers approach the cutter, then use the next piece as a push stick or move to the other side of the table and pull the workpiece through the cut. Repeat the pass on the other side of the bar *(above)*. Prepare an extra bar to help set up the cut in step 3.

2 Cutting rabbets for the glass panes
Install a dado head on your table saw slightly wider than the desired rabbets. The tongue remaining after the rabbets are cut should measure at least ⅜ inch. Install a wooden auxiliary fence and mark the rabbet depth on it—the combined thickness of the glass and the molding strip. Position the auxiliary fence over the dado head, ensuring that the metal fence is clear of the cutters. Turn on the saw and slowly crank up the dado head until it forms a relief cut to the marked line. Turn off the saw and mark the width of the rabbets on the leading end of the glazing bar. Butt one of the marks against the outer blade of the dado head, then position the fence flush against the bar. Use three featherboards to support the workpiece as in step 1, adding a support board to provide extra pressure for the featherboard clamped to the table. (Again in this illustration, the featherboard on the outfeed side of the fence has been removed for clarity.) Feed the bars by hand *(left)* until your fingers approach the featherboards, then use the next workpiece to finish the pass. Complete the cut on the final workpiece by pulling it from the outfeed side of the table.

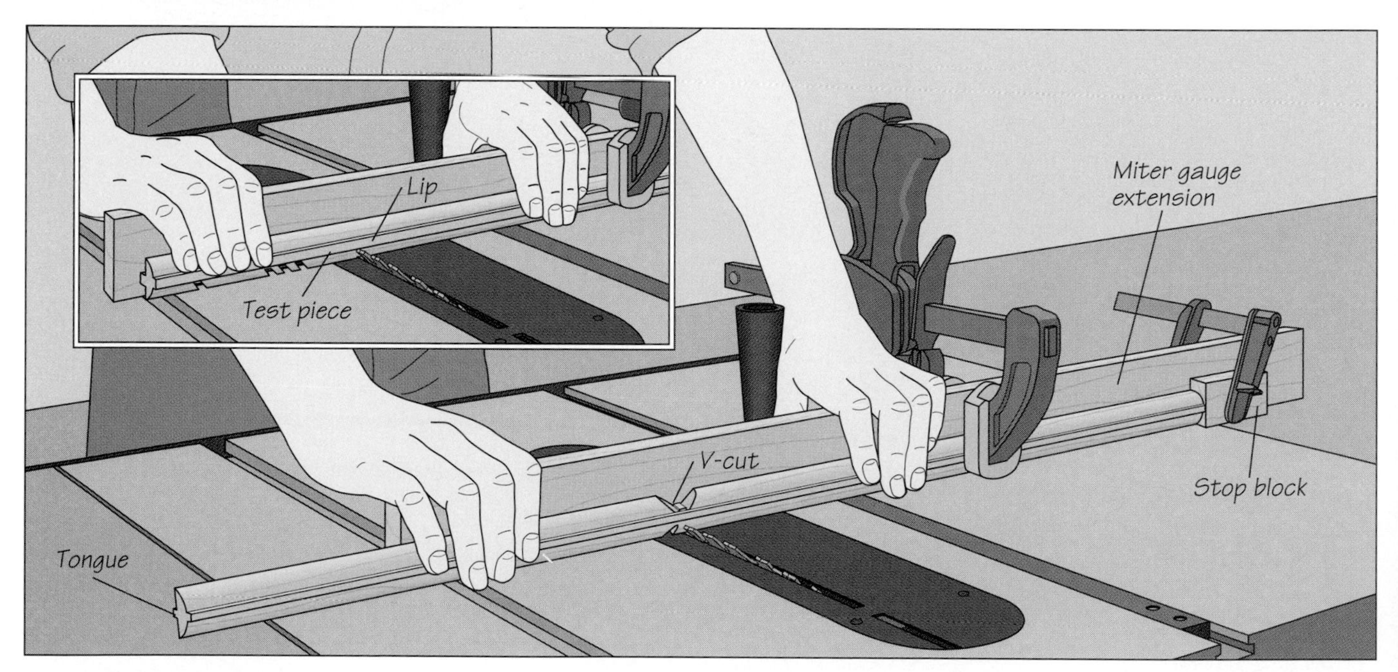

3 Making the miter cuts
Remove the dado head and install a crosscut blade. Adjust the blade angle to 45° and attach a wooden extension to the miter gauge. To set the blade height, hold the extra glazing bar on the saw table so the tongue you cut in step 2 is flush against the extension. The top of the blade should be level with the lower side of the lip. Make a test cut and adjust the blade height until the cutting edge just scores the lip *(inset)*. Then mark the miter cuts on both sides of the bars; at their widest points, the Vs should be the same width as the stock. To make the cut, hold the tongue of the bar flat against the miter gauge extension and align one of the marks with the blade. Butt a stop block against the end of the stock and clamp it to the extension for subsequent cuts. Clamp the workpiece to the extension and feed the glazing bar into the blade while holding it firmly in place. Rotate the piece and make the same cut on the other side of the V. Repeat the process to cut the V on the opposite side of the bar *(above)*.

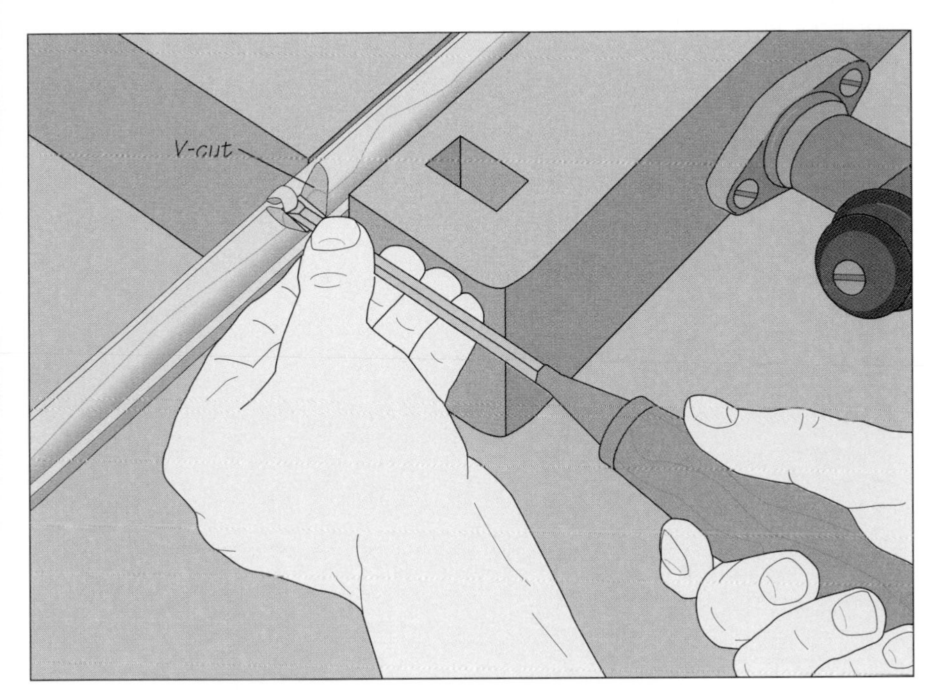

4 Cleaning up the V-cuts
Once all the miter cuts have been made, use a narrow chisel to pare away the waste. The width of the channel at the bottom of the V should equal the width of the lip. Holding the chisel bevel side up, pare away the waste *(left)* until the bottom of the V is smooth and flat. Work carefully to avoid tearout.

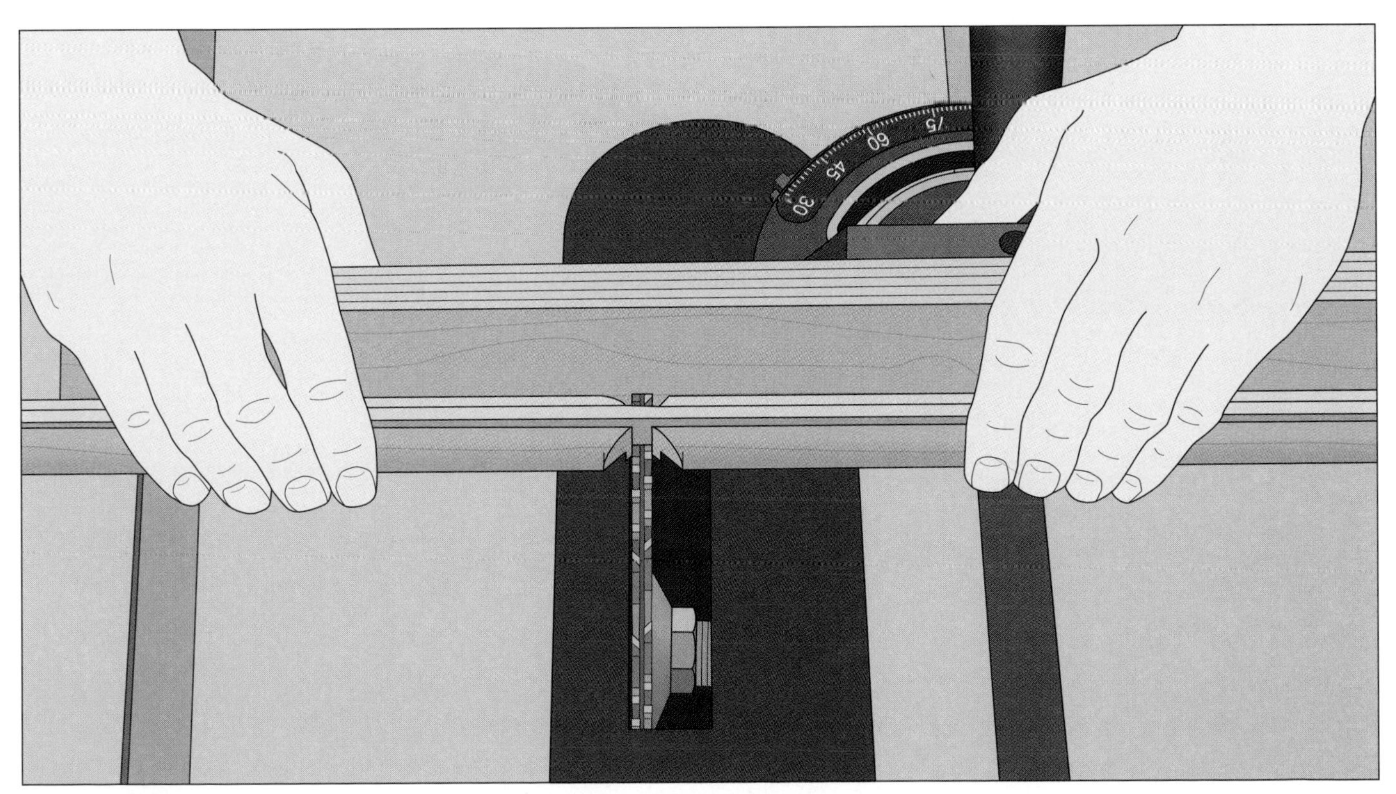

5 Cutting the half-laps
Reinstall the dado head in your table saw and adjust it to the width of the bar's lip. Set the cutting height to one-half the stock thickness. You will be cutting a half-lap in the bottom of one glazing bar, then making an identical cut in the top of the mating piece. Set up the cut by aligning the middle of the V-cut with the dado head, while holding the bar flush against the miter gauge extension. Keep the workpiece flat on the saw table and flush against the extension as you cut the half-laps *(above)*.

DOORS

A length of head casing is fitted over a door and set against a rosette corner block. Framing a door with decorative molding serves an important esthetic function in a room.

A proper door, like a proper chair, may go unnoticed most of the time. It will swing easily, close with a satisfying thud, and rest squarely in its opening. Its style and weight will complement its surroundings. But like a smiling face, a well-built, well-hung, and well-framed door meets friends and strangers with an unspoken, but warm-hearted welcome.

For all its workaday nature, a door and its surroundings are no simple things. The door, frame, and trim must be both sturdy and decorative. This chapter reveals the anatomy of a door and shows just a few of the many styles in use. Starting on page 90, you will see how to build one of the most elegant and popular designs, a frame-and-panel door. Building a door is a challenging task, but the result—a door that is uniquely suited to its settings—is one of woodworking's most gratifying accomplishments.

Doors, of course, serve both interior and exterior uses. Exterior doors have great structural demands placed upon them; the differences in heat and humidity inside and outside alone require that an exterior door be built of heavier material than an interior door, that its design be suited to its use, and that the joinery be of the highest quality. While the environment of an interior door is less harsh, a lifetime of use still demands care in construction.

Once a door is selected or built, it still must be hung. In many ways, hanging a door is the most demanding chore of the finish carpenter. Techniques vary widely, but the job typically comprises several distinct steps, each of which must be tackled with careful attention to detail. Normally, a carpenter will frame the rough opening for a door, which should equal the width of the door plus the thickness of the side jambs and an additional 1-inch space for shims that are used to plumb and straighten the jambs. Door jambs *(page 95)* are made from ¾-inch stock; exterior door jambs should be 1- to 1½-inch thick. Once the jamb is plumb and level in the opening, it is nailed in place. Next, the door is installed with butt hinges mortised into the door edge and the hinge jamb *(page 100)*. For very heavy doors, the hinges are screwed through the jamb into a wall framing member called the trimmer stud. Finally, the casing, or door trim, is installed *(page 112)* to conceal the rough opening and act as a gusset, tying the jamb to the trimmer. Each phase of the door-hanging process is shown in detail in this chapter.

A leaf of a brass butt hinge is fastened to the edge of a frame-and-panel door. Once the other half is screwed to the door jamb, the door can be hung. A pin will hold the hinge leaves together.

ANATOMY OF A DOOR

Doors can be broadly divided into two groups: exterior and interior. While their styles may be similar, the construction differs. Exterior doors are typically 1¾ inches thick and 80 inches high, although in older homes doors are often 82 to 84 inches high. The width varies with location. Front doors are usually 36 inches wide. Back doors and other entry doors can be as narrow as 32 inches. Interior doors range from 24 to 36 inches wide and are typically 80 inches high and 1⅜ inches thick.

Most doors in North America are either frame-and-panel, solid-core, or hollow-core. In frame-and-panel doors, a framework of stiles, rails, and mullions supports solid wood panels that float in grooves milled in the inside edges of the framework. Mortise-and-tenon joints are commonly used to assemble the framework. Cope-and-stick joints are another option, but they must be reinforced with splines to withstand stress.

Solid-core doors consist of a plywood veneer glued over a particleboard core. Hollow-core doors have a lightweight interior, usually cardboard. As shown on page 87, there are many door styles to choose from.

Hanging a door involves building and installing a jamb to fit the rough opening. Rough openings are typically framed by king studs on each side with a trimmer stud attached inside the king studs. A header rests on the trimmers and constitutes the top of the rough opening. When you build your jamb, allow ½ inch of clearance between the jamb and the header and trimmers to allow for shimming. The anatomy below shows a typical rough door opening and a door with the jamb, and the casing installed on one side.

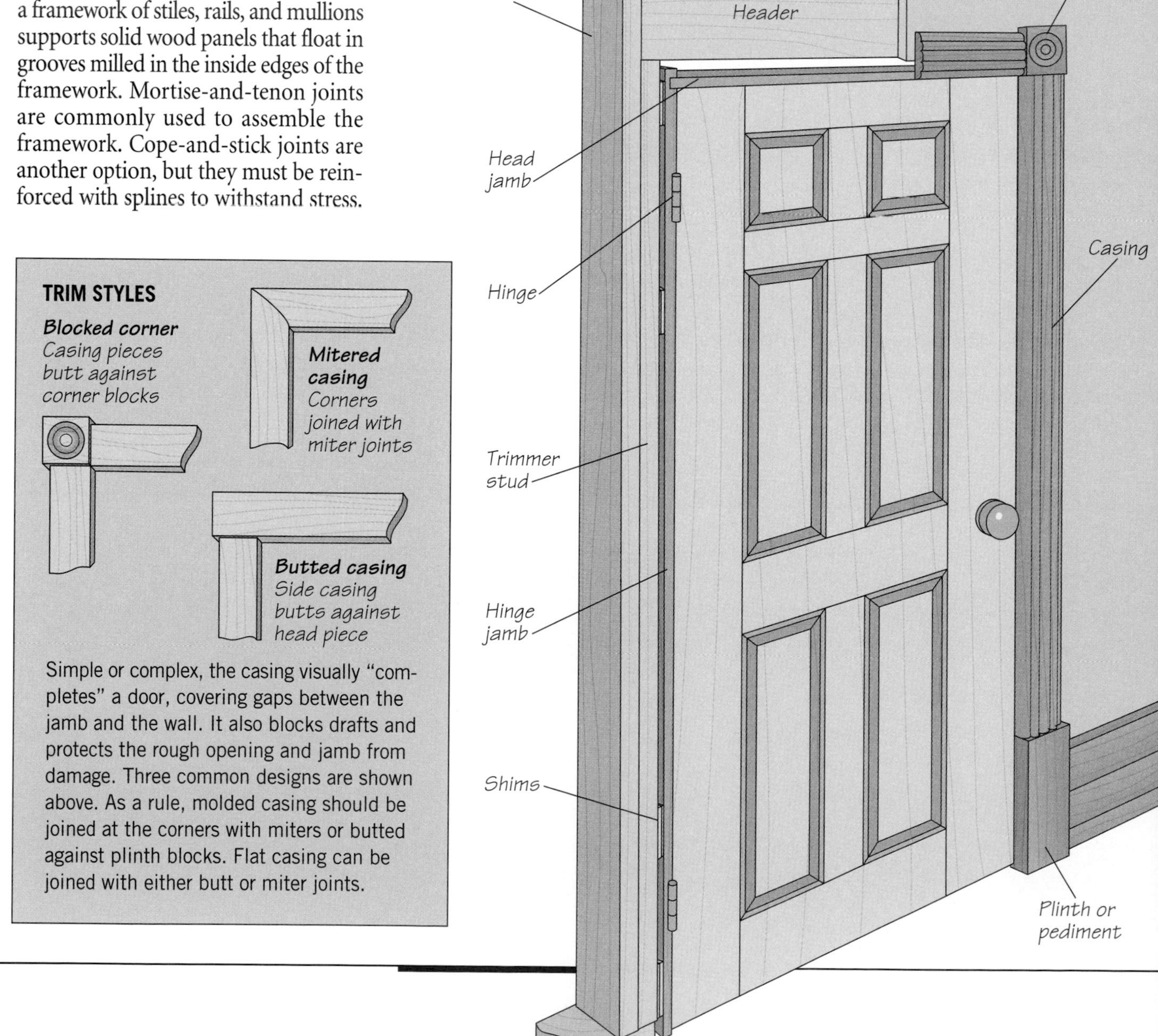

Simple or complex, the casing visually "completes" a door, covering gaps between the jamb and the wall. It also blocks drafts and protects the rough opening and jamb from damage. Three common designs are shown above. As a rule, molded casing should be joined at the corners with miters or butted against plinth blocks. Flat casing can be joined with either butt or miter joints.

DOOR STYLES
French
Suitable for indoor and outdoor use
Solid-core
Suitable for indoor and outdoor use
Frame-and-panel
Suitable for indoor and outdoor use
Glass-and-wood-paneled
Only suitable for outdoor use
Victorian screen
Only suitable for outdoor use
Board-and-batten
Suitable for indoor and outdoor use

TOOLS AND DOOR HARDWARE

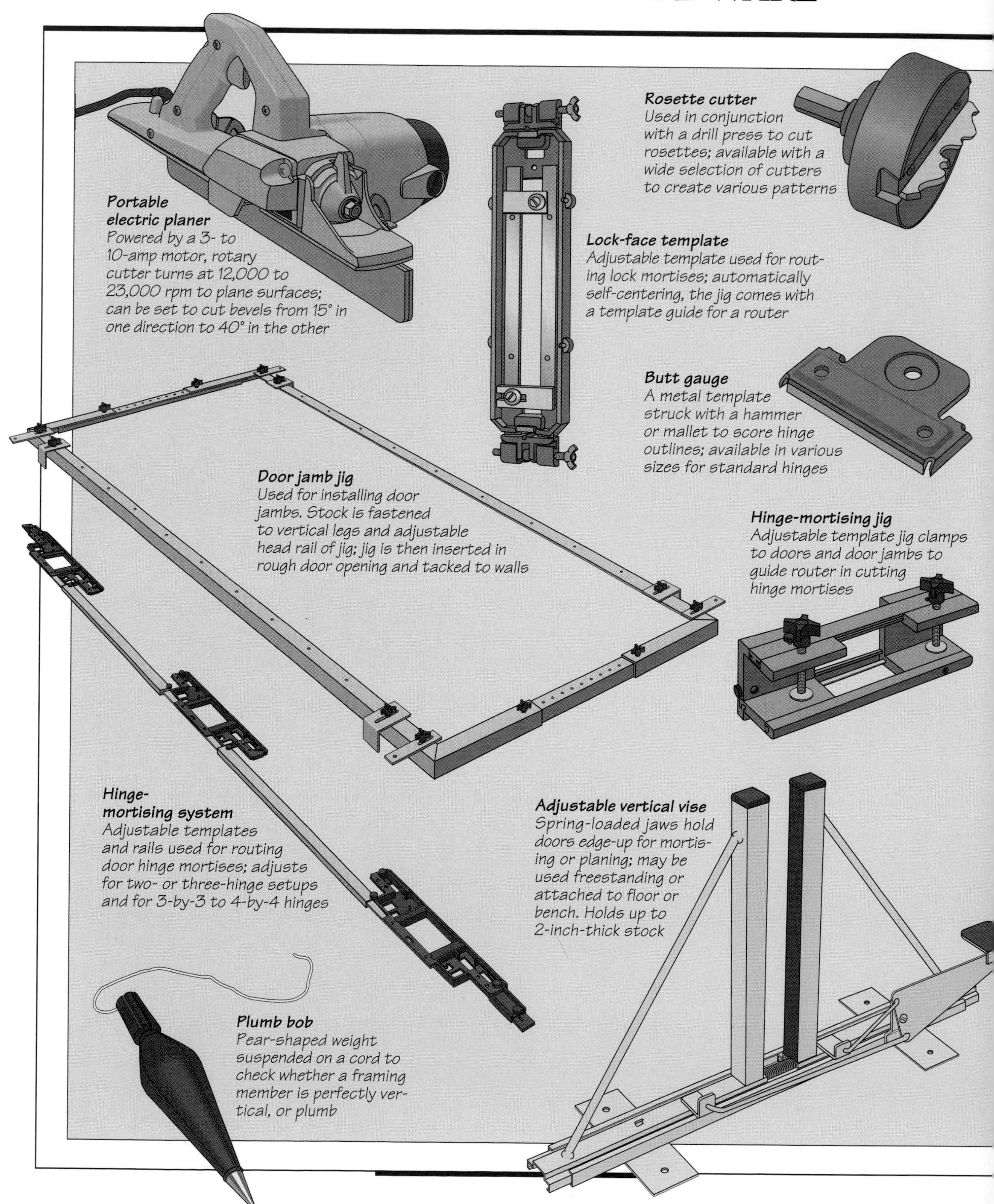

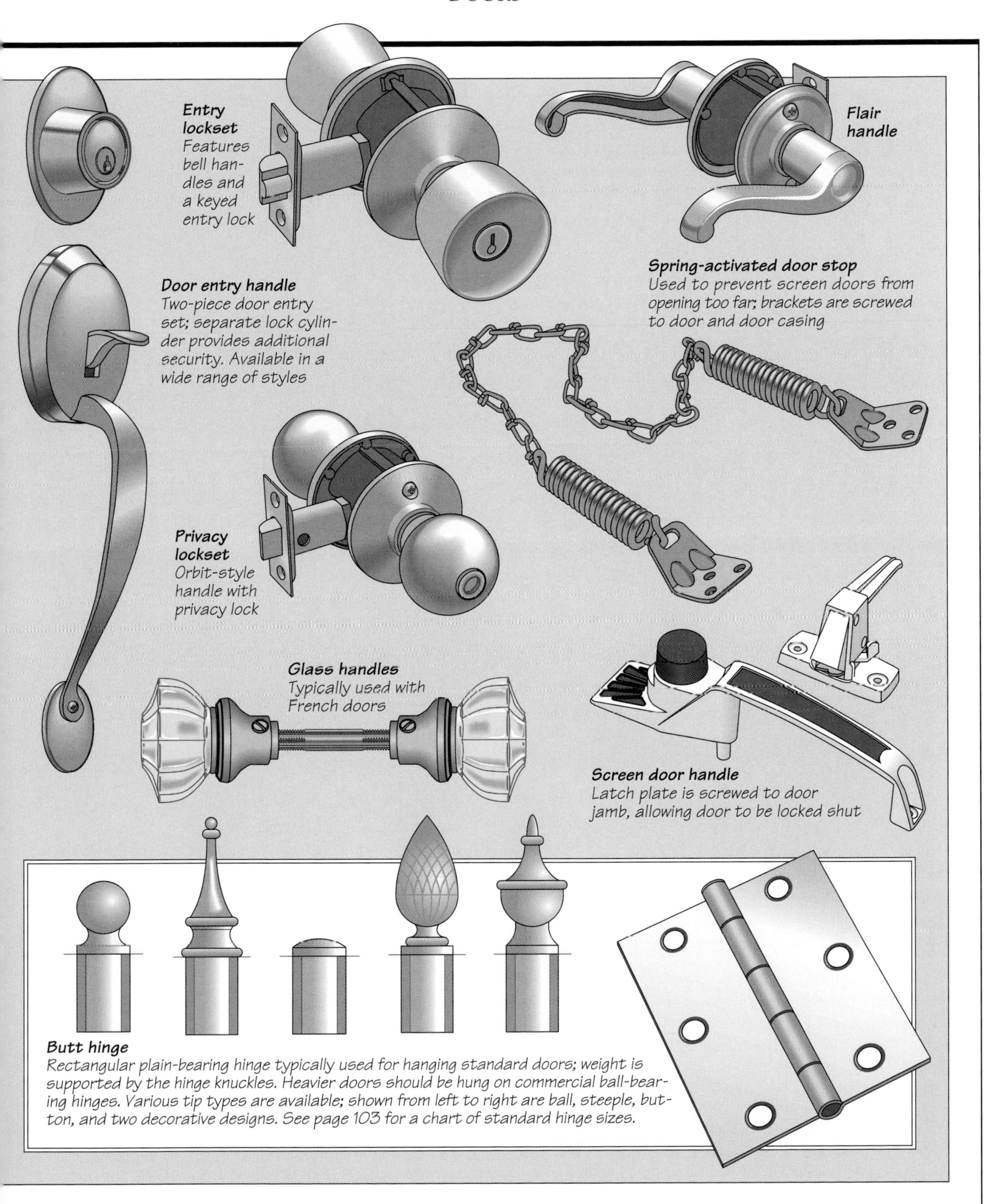
Entry lockset
Features bell handles and a keyed entry lock
Flair handle
Spring-activated door stop
Used to prevent screen doors from opening too far; brackets are screwed to door and door casing
Door entry handle
Two-piece door entry set; separate lock cylinder provides additional security. Available in a wide range of styles
Privacy lockset
Orbit-style handle with privacy lock
Glass handles
Typically used with French doors
Screen door handle
Latch plate is screwed to door jamb, allowing door to be locked shut
Butt hinge
Rectangular plain-bearing hinge typically used for hanging standard doors; weight is supported by the hinge knuckles. Heavier doors should be hung on commercial ball-bearing hinges. Various tip types are available; shown from left to right are ball, steeple, button, and two decorative designs. See page 103 for a chart of standard hinge sizes.

A shaper is invaluable for making frame-and-panel doors. Fitted with cope-and-stick cutter sets, it will prepare the stiles and rails for assembly, cutting grooves for the "floating" panels that fill the frame and carving a decorative molding along the inside edges of the frame at the same time. Then, equipped with a panel-raising bit, the shaper can form bevels on the panel edges, as shown in the photo at right. The large shop-made featherboard clamped to the shaper's fence protects the user from the cutter and holds the panel flat on the table. Step-by-step instructions for building a six-panel Federal-style door are provided below and on the following pages.

MAKING A FRAME-AND-PANEL DOOR

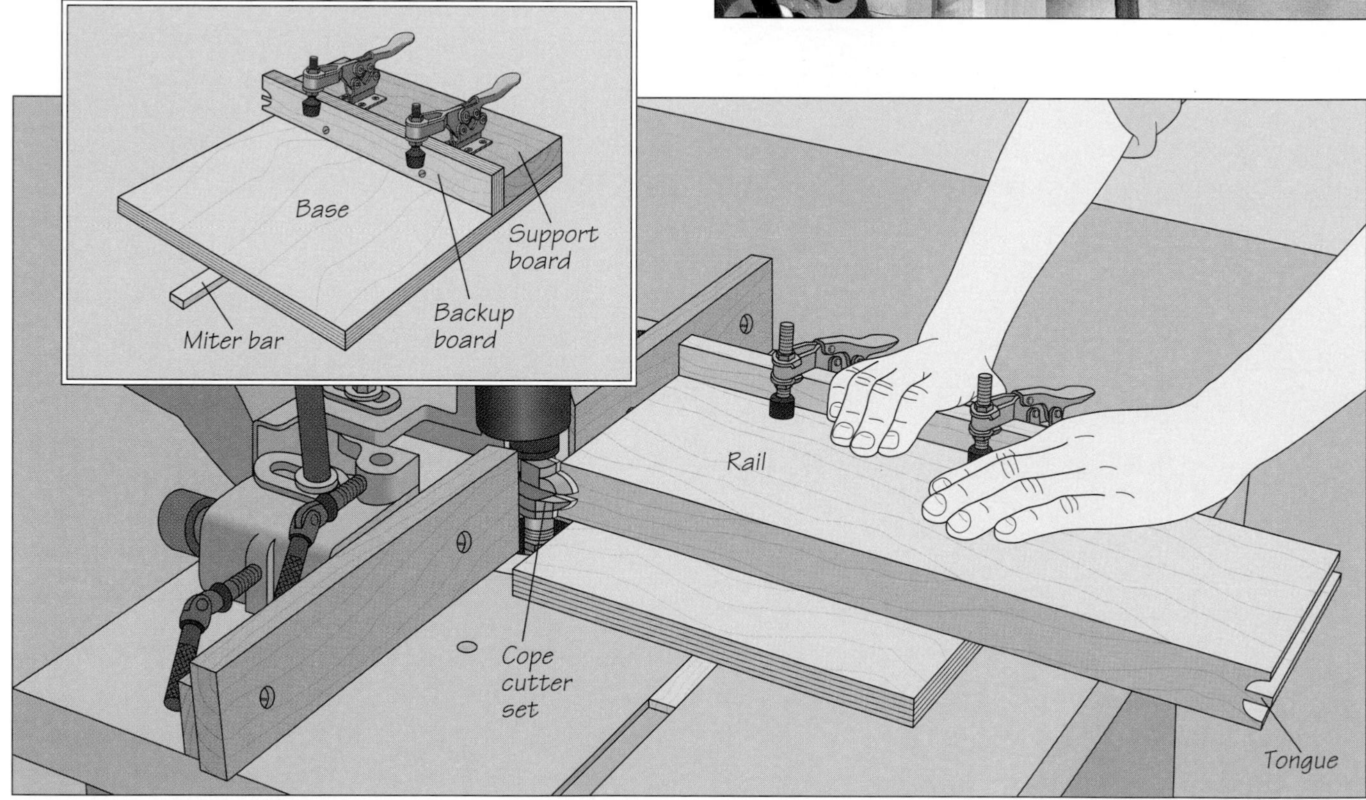

1 Making cope cuts on the rails

As shown on page 86, a six-panel door features two stiles, a top and bottom rail, two median rails, and three mullions. Cut your stock to size, then install a coping cutter set and guard on the shaper. To feed the rails, build the coping jig shown in the inset. The jig consists of a plywood base, a miter bar screwed to the underside of the base, a 2-by-4 support board fastened flush with the back edge of the base, and a plywood backup board screwed to the support board. To prevent tearout on the rails, the backup board should support the workpiece for the full width of cut. Screw two toggle clamps to the support board. Next, mark the tongue location on one of the rails, centered on the edge of the board. Position the jig on the shaper table, set the one of the rails on the jig and adjust the cutter height to align the cutter with the tongue mark. Then clamp the rail to the jig, aligning the board end with the end of the backup board so the cutter will shape the entire edge. Now make the cut, pushing the jig across the table. Repeat the cut on the other end of the rail *(above)*, then make the cuts on both ends of the remaining rails and mullions.

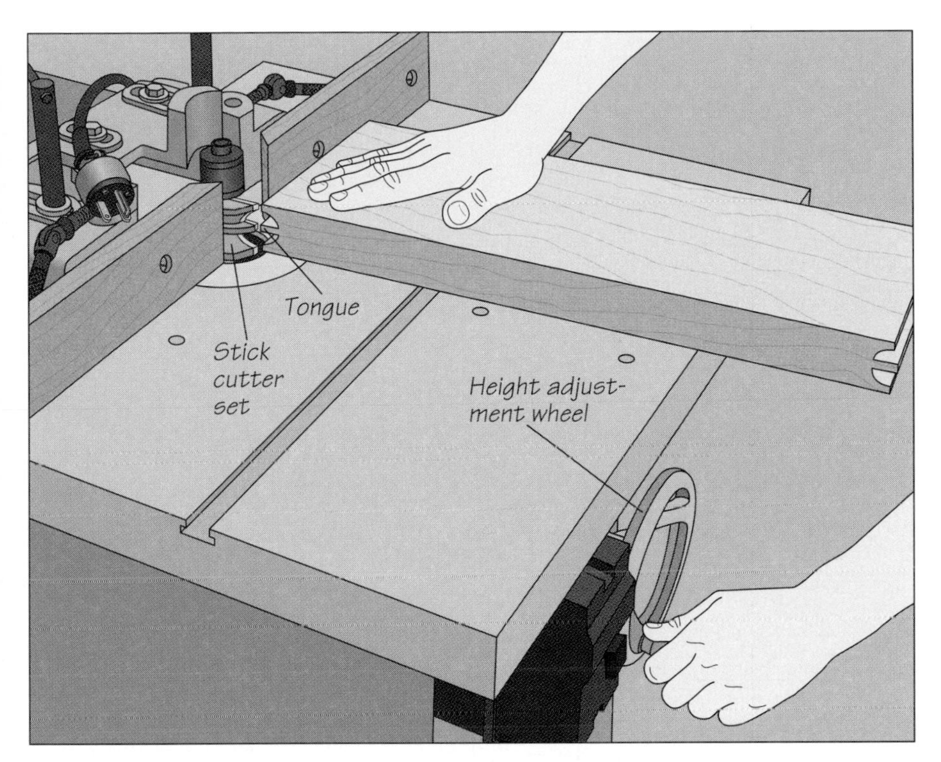

2 Adjusting the stick cutter

Once all the cope cuts are made, replace the cope cutter with the matching stick cutter set. This setup will shape the edges of the stiles, mullions, and rails with a decorative profile while cutting grooves to accommodate the tongues and panels. To set the cutting height, butt the end of one of the coped rails against the stick cutter, then adjust the height of the spindle so the groove cutter is level with the tongue on the rail *(left)*.

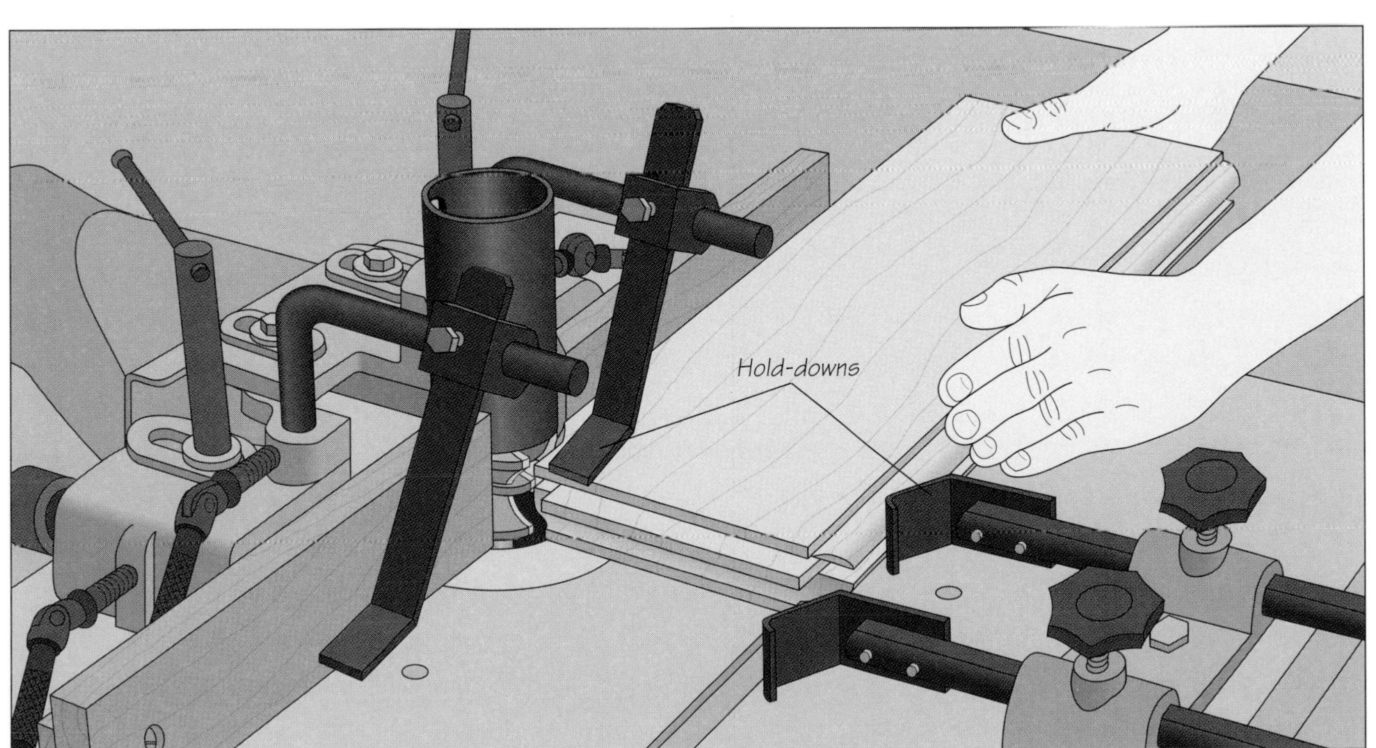

3 Making the stick cuts

Adjust the fence to shape the entire edge of the stock. Also install commercial or shop-made hold-downs on the fence and shaper table to secure the stock through the cuts and prevent kickback. Shape both edges of the median rails and mullions, feeding the stock across the table with both hands *(above)*, but shape only the inside edges of the stiles and top and bottom rails.

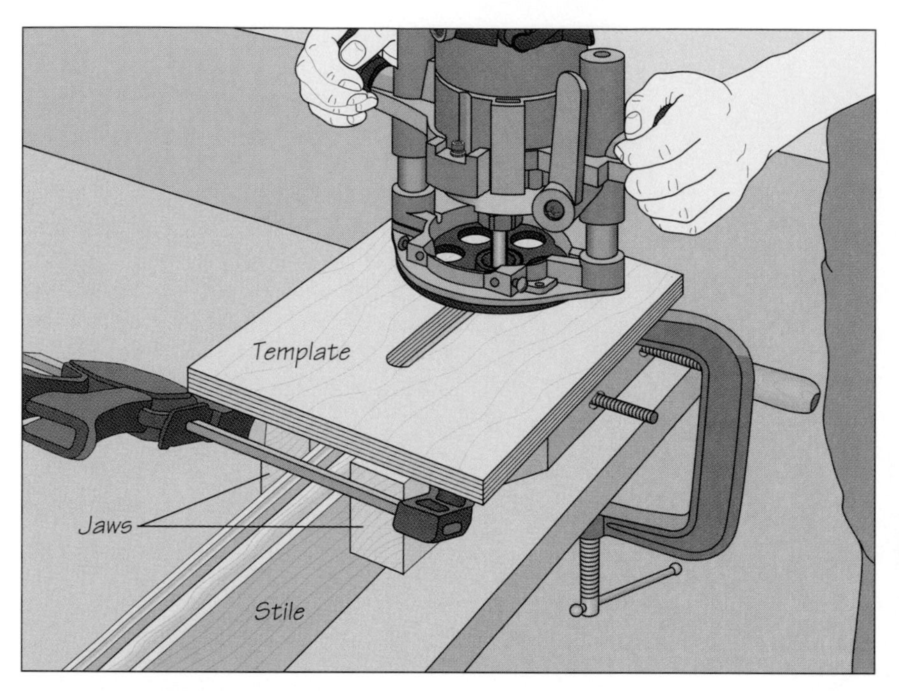

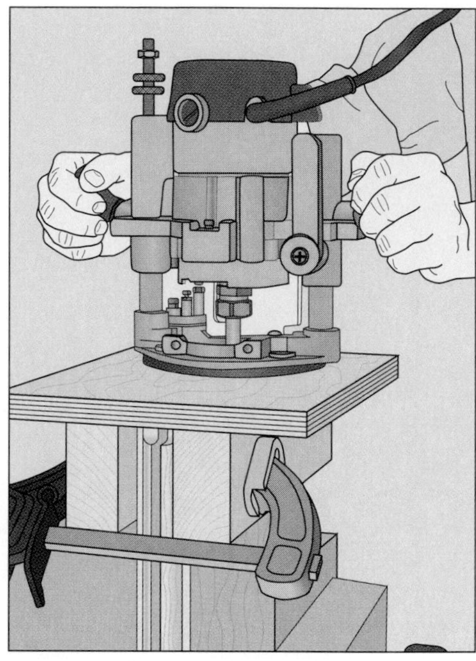

4 Routing mortises in the stiles and rails

Reinforce the joints between the stiles and rails with splines. To determine their locations, test-assemble the stiles and rails and mark the center of the joints between them. Take the assembly apart and secure a stile edge-up on a work surface. Use a router fitted with a mortising bit and a template guide to cut mortises for the splines. To guide the tool, build the jig shown above, made from a piece of ¾-inch plywood with a slot in the middle and two 2-by-4 jaws screwed to the bottom of the template to straddle the stile. The slot should be the size of the groove you wish to cut plus the diameter of the template guide you will attach to the router. Clamp the jig to the stile, then set the cutting depth to cut a 1½-inch-deep mortise in the stile. Turn on the router and make the cut, guiding the template guide along the inside edges of the jig slot *(above, left)*. Repeat the cut at the other end of the stile, at both ends of the other stile, and at the center of median rails. Next, secure the rails end up and rout grooves in their ends the same way *(above, right)*.

5 Test-fitting the joint

Once all the grooves are cut, make splines that fit the mortises and are shorter than the combined depth of two mortises. The grain of the splines should run in the same direction as the rails. Test fit one of the joints before glue-up *(right)*. The joint should fit together smoothly without binding. If the fit is too tight, trim the spline and test-fit the joint again. Finally, make reference marks on all the rails and stiles to help you assemble them properly during glue-up *(page 94)*.

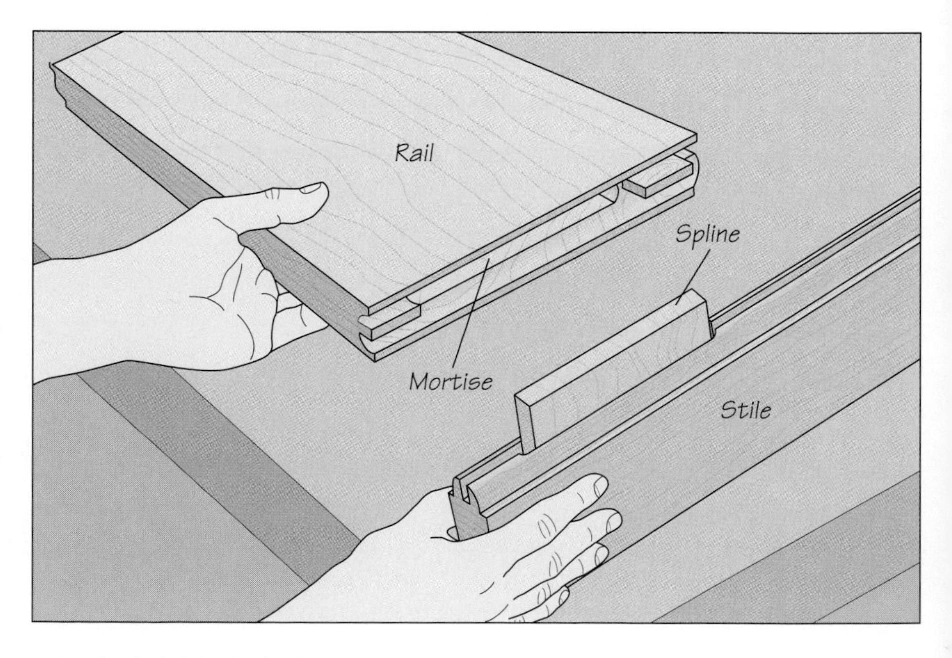

6 Raising the panels
To help you size the panels, assemble the door stiles, mullions, and rails and measure the openings. Add ½ inch to each dimension to allow for the ¼ inch along the edge of the panel that will fit into the grooves. Cut the panels to size; your stock should be no thicker than the stock used for the stiles and rails. Install a panel-raising bit and matching rub bearing in the shaper, and adjust the fence even with the rub bearing. Then adjust the cutter height so the raised edges of the panels will penetrate the grooves by ¼ inch when the panel is cut on both sides. Clamp a wide featherboard to the shaper fence to shield you from the cutter and hold the panel flat on the shaper table. Feed each panel face-up into the cutter, using your left hand to keep the workpiece flush against the fence *(right)*. To prevent tearout, shape the panel ends first, and then the sides. Once one side of the panel has been shaped, turn it end-for-end and repeat on the other edge. Then turn the panel over and repeat the series of cuts.

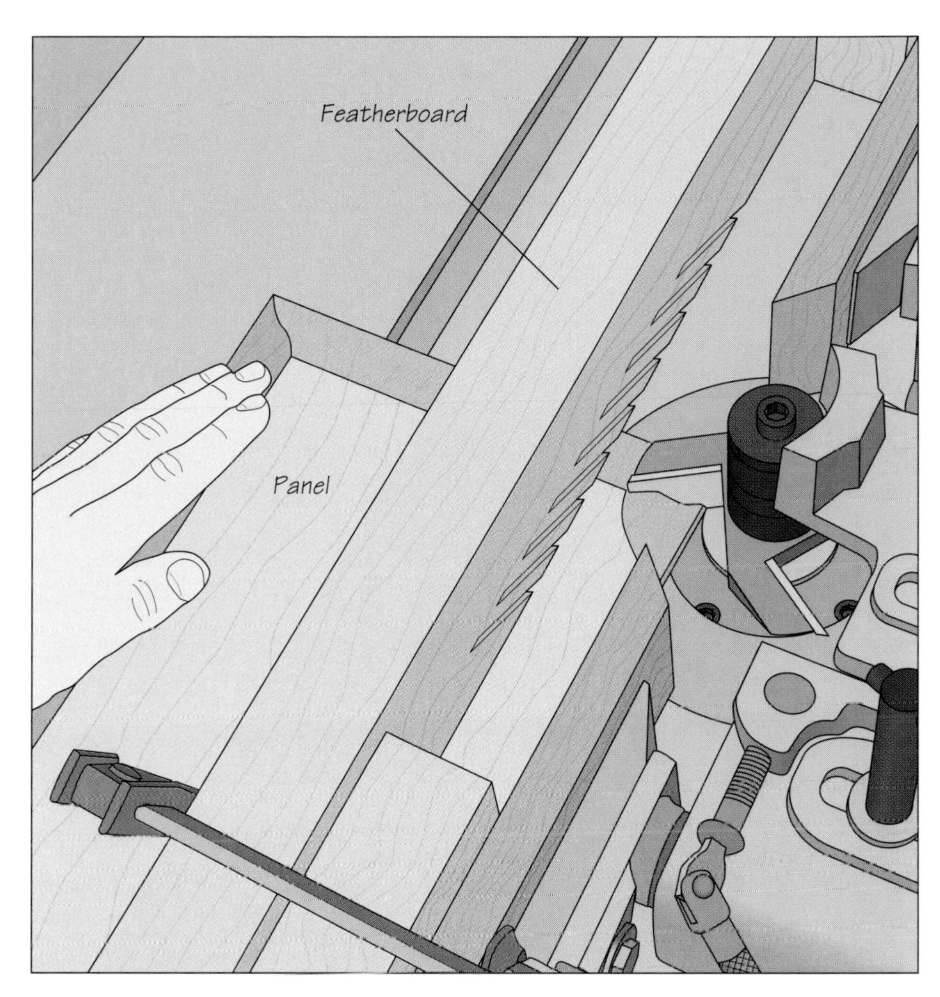

7 Testing the fit of the panels
Once you have shaped the first panel, fit it into one of the grooves in a stile *(left)*. The pieces should fit together snugly, with the panel extending ¼ inch into the groove. If not, adjust the cutting height, repeat the cuts and test the fit again. Once you are satisfied with the fit, raise the remaining panels.

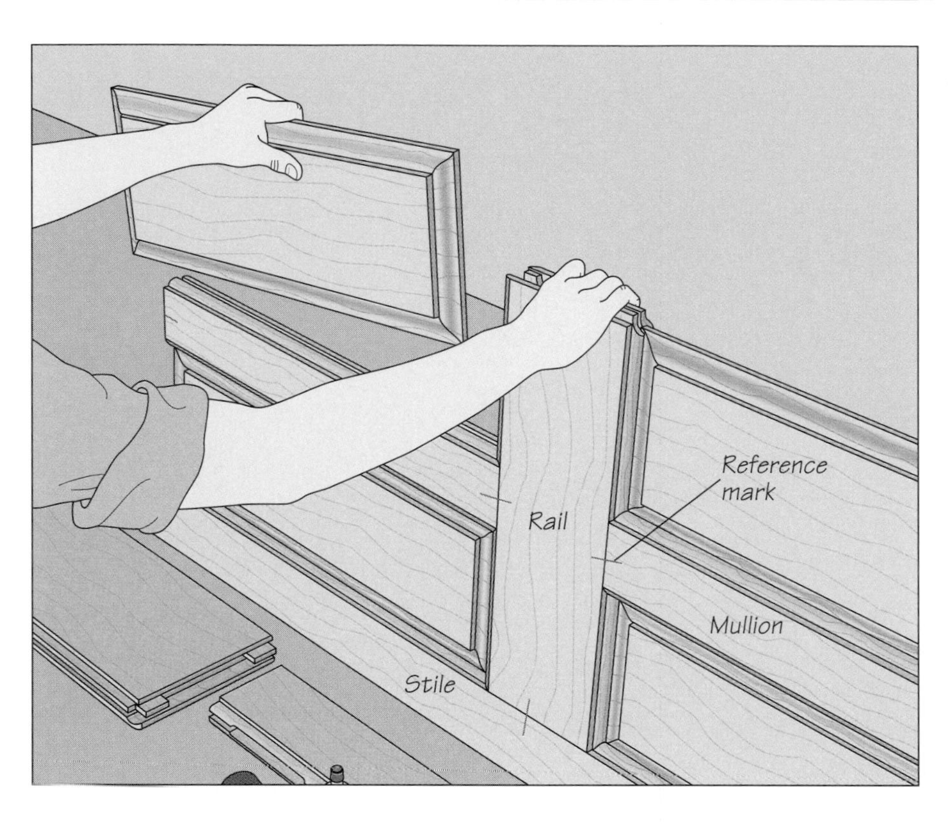

8 Assembling the door

Lay out all the pieces of the door close at hand so that you can assemble the door quickly before the glue begins to set. Start building the door by setting a stile edge-up on the floor. Apply glue in the mortises in the stile and its adjoining rails as well as on the splines. Do not spread any glue in the panel grooves. Insert the splines in the stile mortises and fit the rails in place. Use the reference marks you made earlier to help you assemble the pieces properly. Tap the top ends of the rails lightly with a mallet to close the joints. Now, seat panels between the stiles and rails. Continue in this fashion, applying glue in the spline grooves and on the splines and fitting the pieces in place *(left)* until the door is assembled.

9 Clamping the door

Lay four bar clamps on the floor, one for each rail. Carefully lay the assembled door on the clamps so the bars of the clamps are aligned with the rails. To protect your stock, place wood pads the length of the door between the clamp jaws and the door edges. Tighten the clamps just enough to close the joints. Then clamp the door from top to bottom along the mullions. If you do not have a clamp that is long enough to span the door, use two clamps, positioning them so that their tailstops contact each other near the middle of the door. Use shorter wood pads to protect the door from these clamp jaws. Install three more clamps across the top face of the door, aligning the bars with the top, bottom, and middle rail. Finish tightening all the clamps until glue squeezes out of the joints *(right)*. Then use a try square to check that the corners of the door are square; adjust the clamping pressure, if necessary. Once the glue has dried, use a paint scraper to remove any remaining adhesive. When the glue has cured, sand and finish the door.

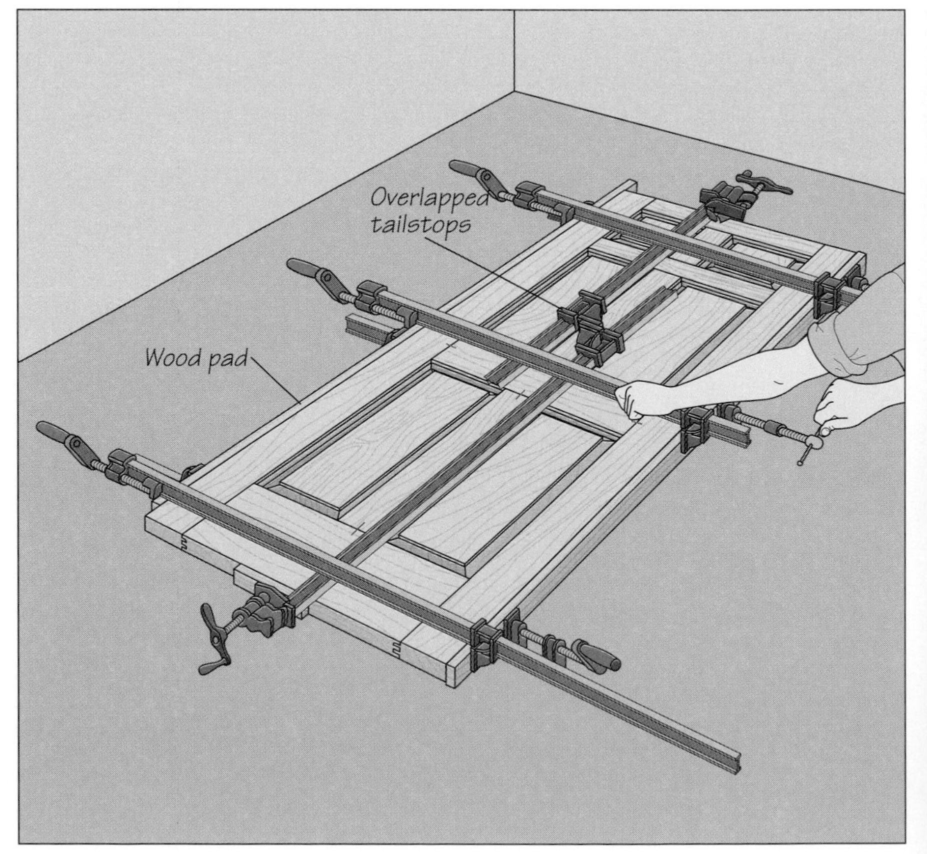

DOOR JAMBS

If you will be hanging several doors, a commercial door jamb jig could prove to be a worthwhile purchase. Its framework of metal legs and rails will keep a jamb square and hold it in position in the rough opening while you set it level and plumb, and fasten it to the trimmer stud.

MAKING AND INSTALLING A DOOR JAMB

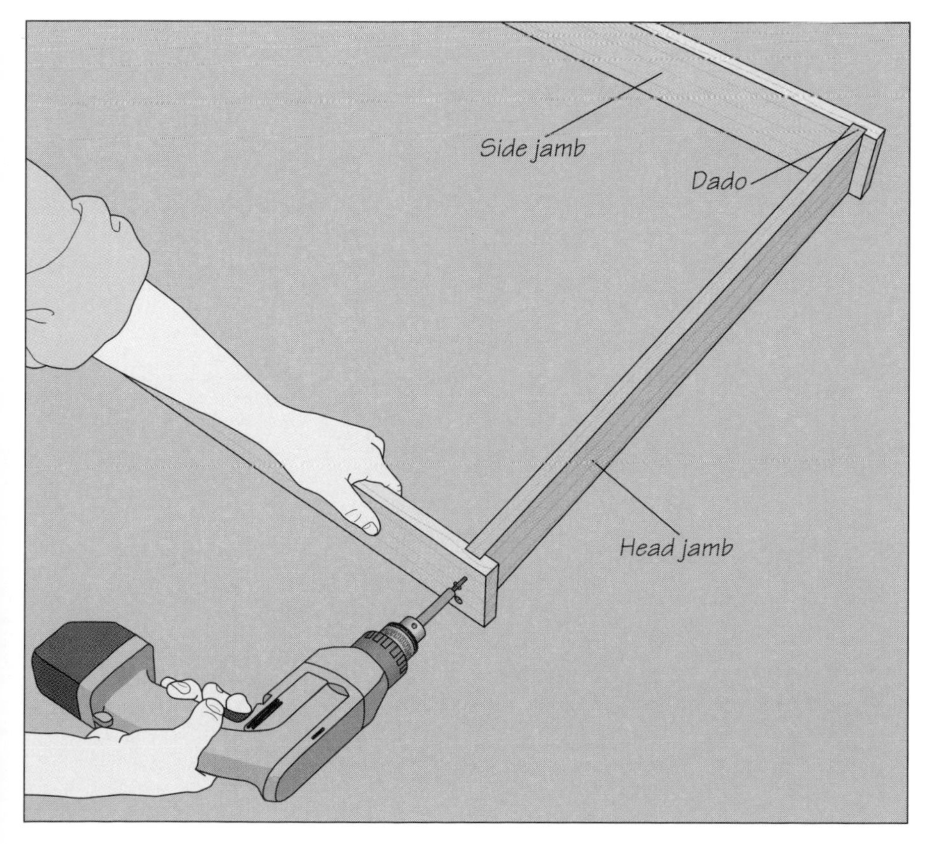

1 Building the jamb
Select straight-grained ¾-inch-thick stock for your jamb. Rip the stock as wide as the thickness of the wall, then cut the side and head jambs to length. Make the side jambs slightly longer than the height of the door so they extend roughly from the floor to the header. Trim the head jamb to the width of the door plus 5⁄32 inch to allow 3⁄32 inch of clearance on the door's latch side and 1⁄16 inch on its hinge edge. If you will use dado joints to join the head and side jambs, as shown at left, add the depth of the dadoes to the length of the head jamb. Once you have cut the dadoes, fit the end of the head jamb into one of the side jambs and screw the pieces together. Repeat for the other side jamb. Finally, cut a spacer, or spreader, to the width of the jamb's opening. This board will be set on the floor between the side jambs to keep the assembly square as it is being installed.

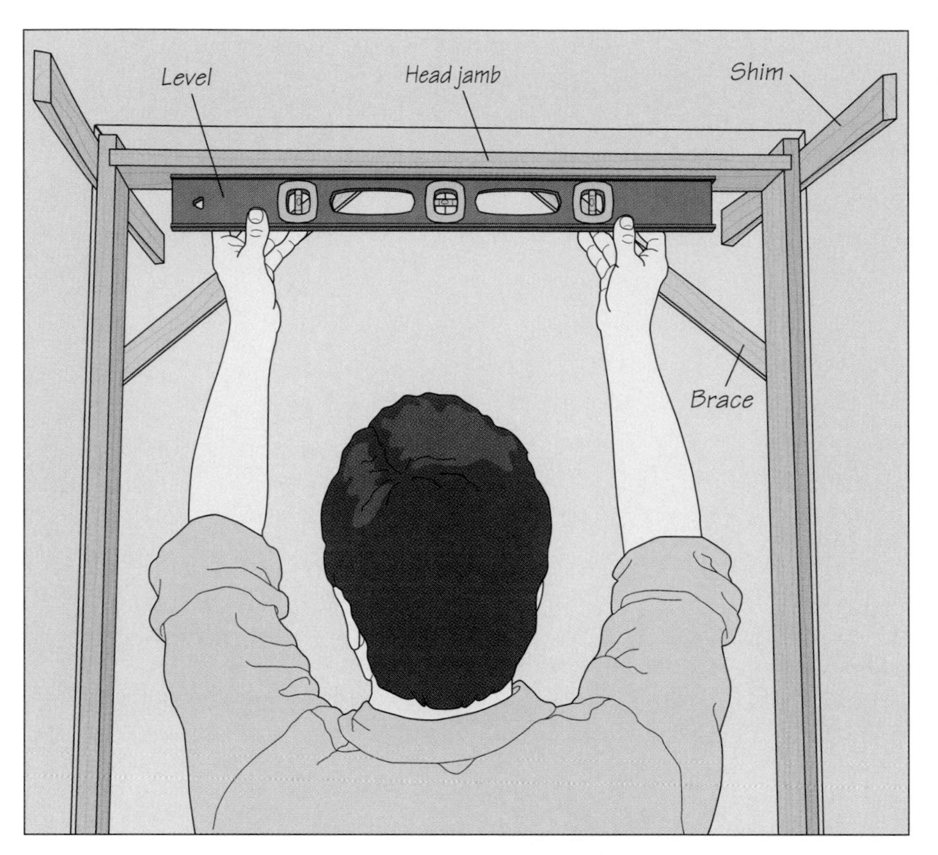

2 Setting the jamb

Tack a bracing board diagonally across each top corner of the door opening to keep the edges of the jamb flush with the walls. Position the jamb in the opening, butting it against the braces, and place the spreader on the floor between the side jambs. Tap shims between the side jambs at both ends of the head jamb to center the assembly in the opening; insert the shims in pairs from opposite sides of the jamb. (Shims are tapered wedges of wood that are usually sold in small bundles at hardware stores and lumberyards.) Then use a carpenter's level to check the head jamb for level *(left)* and shift the assembly slightly if necessary. In the process, one of the side jambs may be raised off the floor. If so, measure the gap and trim the opposite side jamb by the same amount. Reposition the jamb in the opening, centering and leveling it again. Both side jambs will now be on the floor. Nail the jamb to the rough opening through the shims into the trimmer studs. Set the nail heads.

3 Checking for plumb

Tap shims between the side jambs and the wall at both ends of the spreader. Then check the jamb for plumb. To set up the plumb bob, mark the center of the head jamb at one edge, then transfer your mark to the spreader. Tack a small finishing nail into the edge of the head jamb so the plumb bob cord will hang directly below the center mark. Suspend the bob from the nail so the point of the bob hangs just above the spreader. Tap the shims beside the spreader in or out to align the center mark directly under the bob. Drive finishing nails through the side jambs and the shims to secure the side jambs at the bottom of the opening.

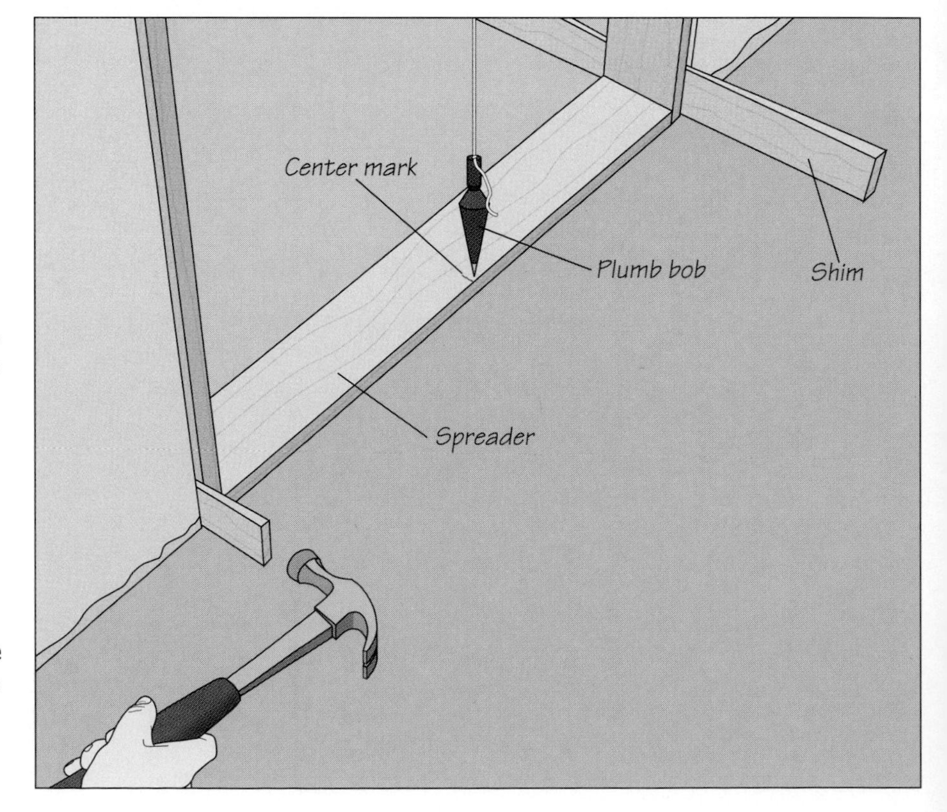

4 Squaring the side jambs
Insert three additional pairs of shims behind each side jamb, positioning them at the hinge and lock strike plate locations. The shims should be wedged in tightly. Although the side jambs are plumb, they may be slightly bowed from top to bottom. To ensure the jambs are straight, press a straightedge against the jamb to flatten it as you nail through the shims *(right)*. In this case, straight 1-by-4 stock is used; a 6-foot-long carpenter's level will also work fine.

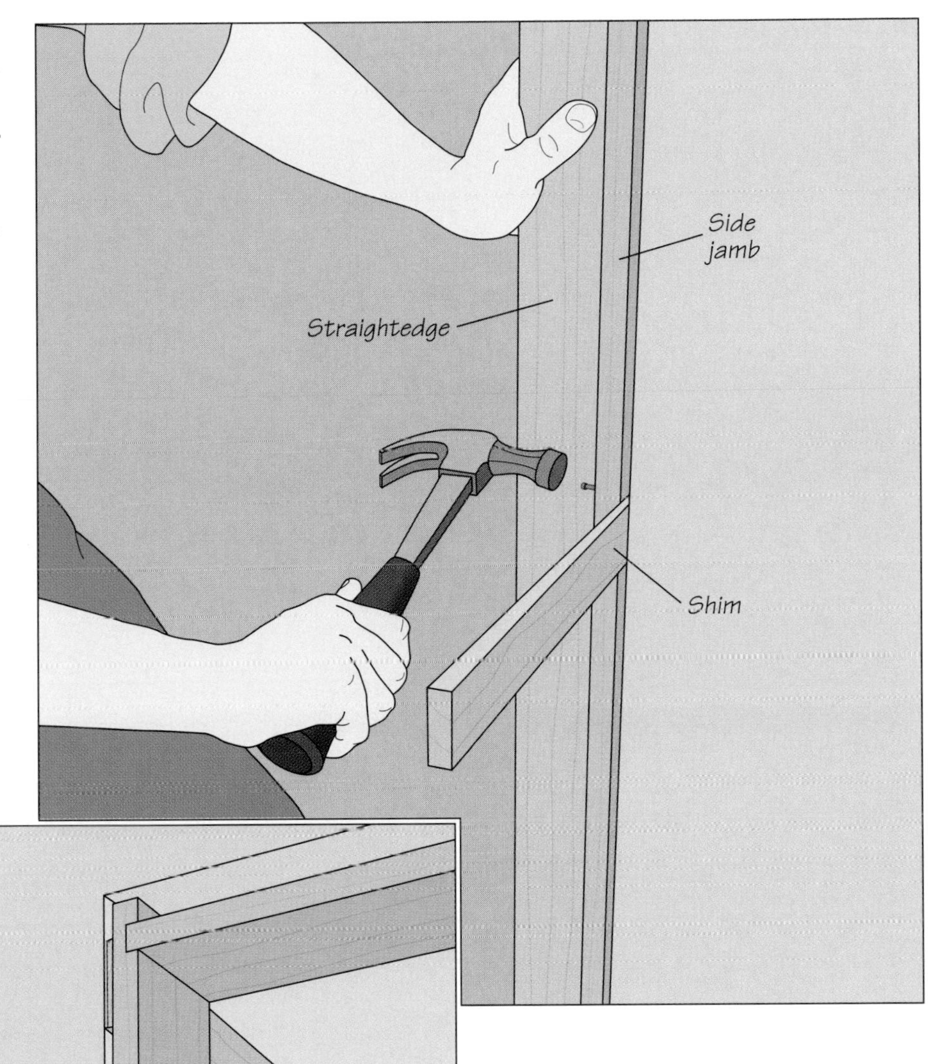

5 Trimming the shims
Once you have shimmed the jamb, cut the shims off flush with the wall using a utility knife. Hold the end of the shim and slice across it repeatedly *(left)* until the waste piece can be broken off easily.

INSTALLING DOORSTOPS

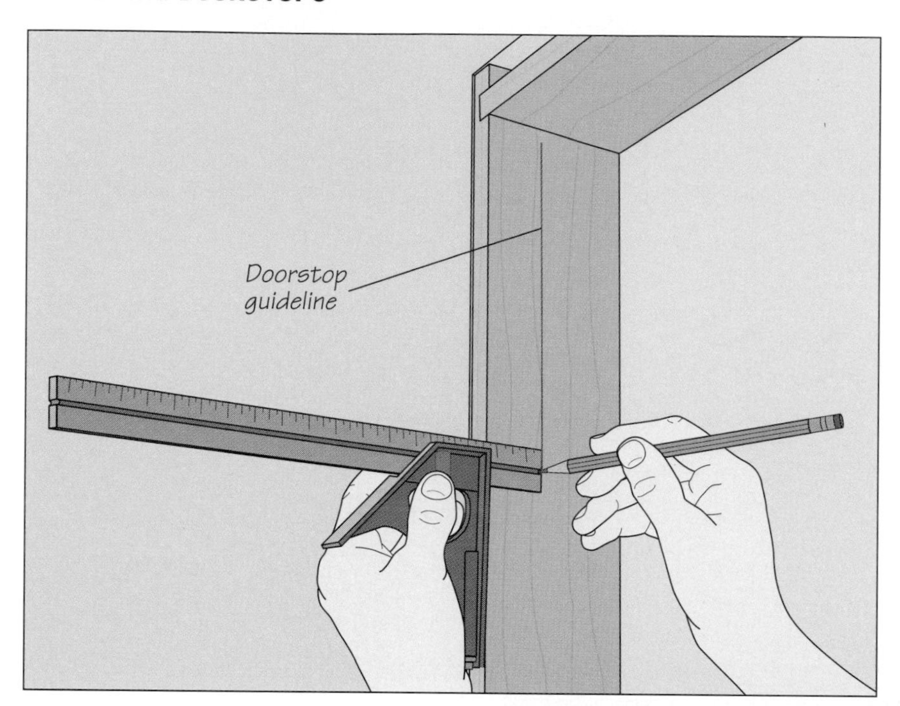

1 Laying out the doorstops
Doorstops can be installed after the door is in place or, as shown at left, once the jamb is installed. Mark a guideline for the doorstops with a combination square and a pencil. Adjust the combination square to the thickness of the door and butt the handle against the door-opening edge of the latch jamb. Starting at the top of the jamb, hold the pencil against the end of the ruler and run the square down the jamb to mark the line *(left).* Before marking the hinge jamb, add 1⁄16 inch to allow clearance for the hinges and prevent the door from binding when it is closed.

2 Preparing the doorstops
You can use either flat or molded stock for the doorstops. The pieces can be joined at the corners with butt joints, miters, or coped joints. In this example, flat stock is being joined with miter joints cut on a chop saw; you can also use a miter box and cut the pieces with a handsaw. Adjust the saw for a 45° cut and butt the first piece against the fence. Clamp a guide board to the saw table to secure the stock flush against the fence and make the cut *(right).* Miter both ends of the head-jamb doorstop and the top end of the side-jamb doorstops. If you are using molded stock, make sure the flat edge will butt against the door when it is closed.

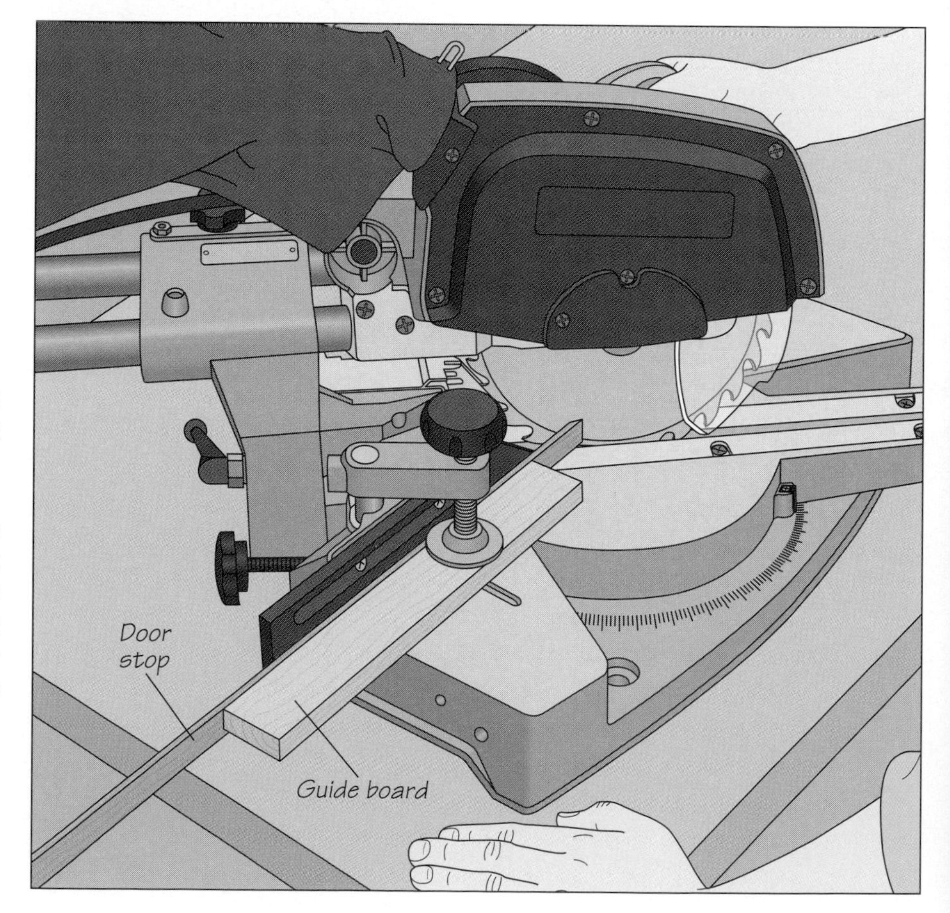

3 Installing the head-jamb doorstop
Once all the doorstops are cut, install them on the jamb. Start with the head-jamb doorstop. Align the edge of the piece with the lines marked on the side jambs and tack it in place with finishing nails *(right)*. Make sure the mitered ends are facing down. Do not drive the nails flush, as you may have to reposition the doorstops once the door is installed. The head-jamb piece will be slightly askew because of the $\frac{1}{16}$-inch offset between the two guidelines on the side jambs.

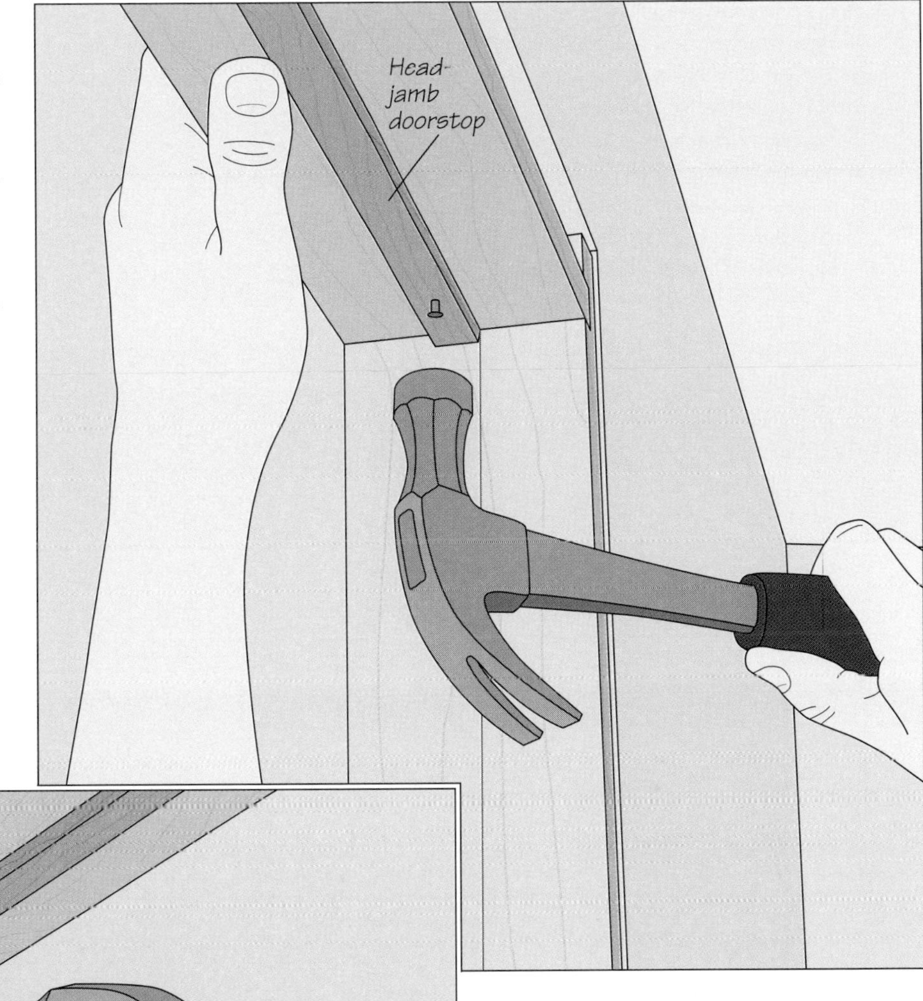

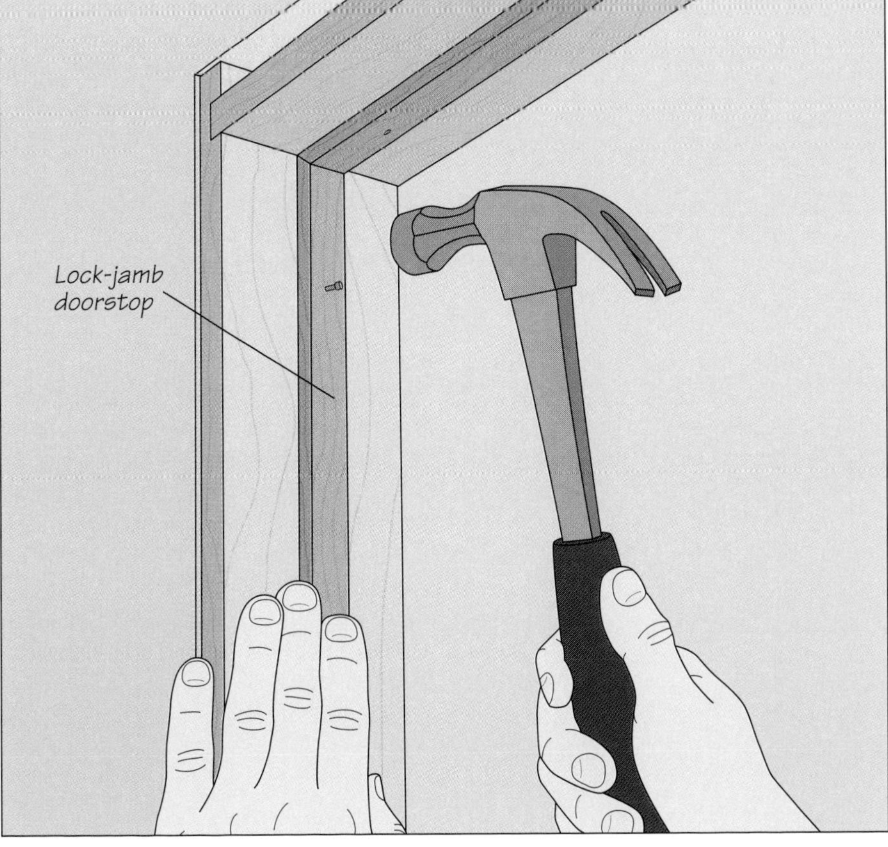

4 Installing the lock-jamb doorstop
Align the doorstop with the guideline on the jamb, butt its mitered end against the end of the head-jamb doorstop and tack the piece in place *(left)*. Make sure the miter joint is tight. Do not install the hinge-jamb stop until you have cut mortises for the hinges and hung the door *(page 100)*.

A chisel pares away the waste from a hinge mortise in a door jamb. By using a chisel that is the same width as the mortise, you can tap the chisel with a mallet to score a series of cuts across the mortise and around its outline. Then push the chisel bevel-side up to shear off the waste wood and clean up the bottom of the cavity.

CUTTING THE HINGE MORTISES ON THE JAMB BY HAND

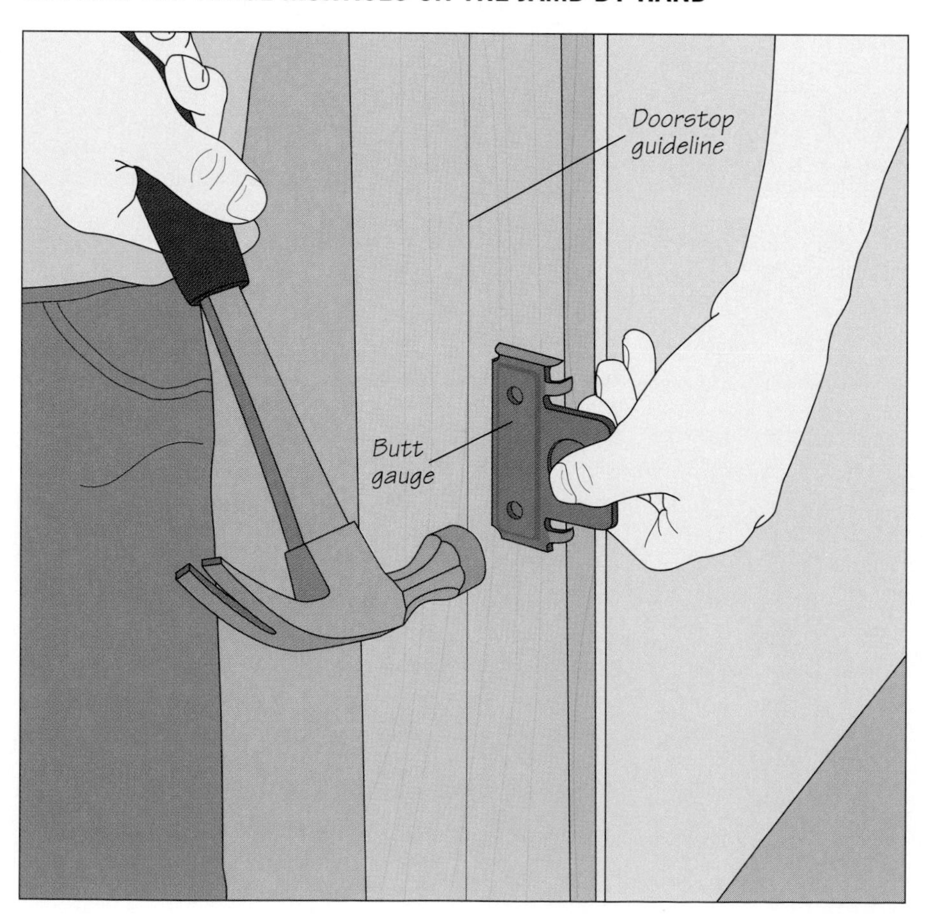

Chiseling out the hinge mortises
There are no firm rules for locating door hinges, but on an 80-inch interior door, they are typically positioned 7 inches from the top and 11 inches from the bottom of the door. If you choose to use a third hinge, locate it midway between the other two. Mark the location of the hinges on the hinge jamb, allowing for ⅛ inch of clearance between the door and the head jamb. You can use a router and a shop-made template jig to cut the hinge mortises *(page 101)*, or a commercial hinge mortising system *(page 102)*. To do the job by hand, you can use a butt gauge to score the hinge mortise outlines on the jamb. Aligning the gauge stops sets the device automatically in place directly over the bottom hinge mark. Now strike its face with a hammer. Repeat to outline the remaining mortises on the jamb. Then use a chisel to cut the mortise, as shown in the photo above.

ROUTING THE HINGE MORTISES ON THE JAMB

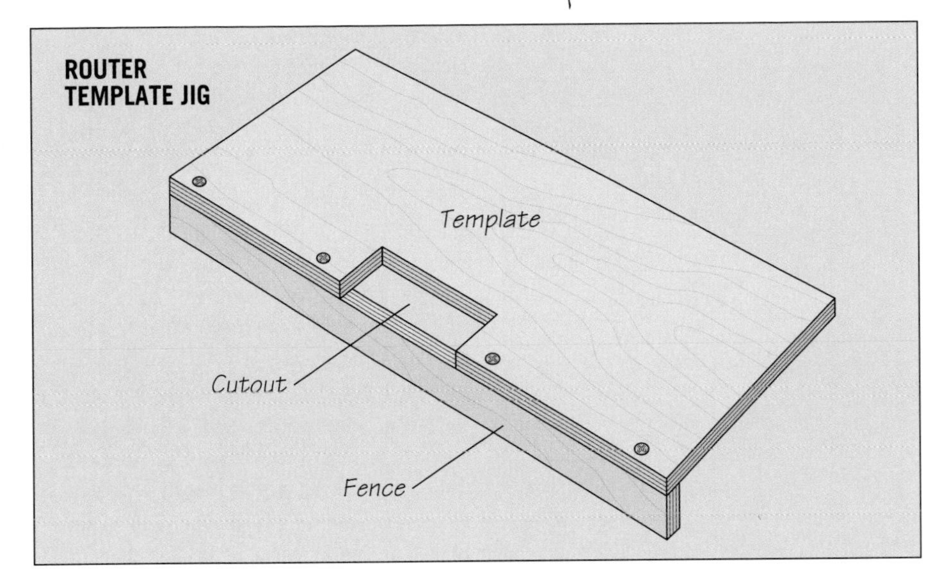

1 Making a router template jig
To rout out hinge mortises on a door jamb, use the template jig shown at left with your router, a straight bit, and a template guide. Make the jig from two pieces of ¾-inch plywood; the template should be wide enough to support the router. Outline the hinge leaf on the template and cut it out, compensating for the template guide and adding the thickness of the fence. Fasten the fence to the template, countersinking the screws.

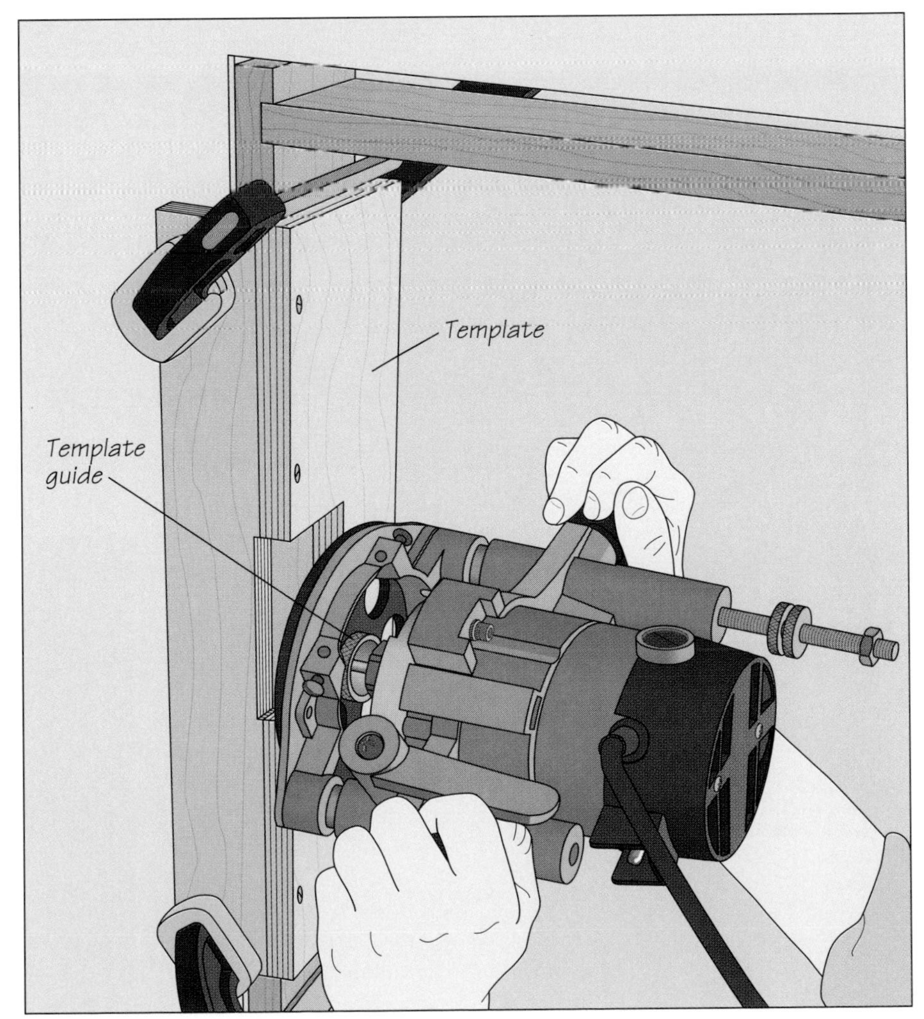

2 Routing the mortises
Mark the hinge locations on the door jamb *(page 100)* and clamp the jig to the jamb, aligning the cutout with one of the marks. Butt the jig fence against the door-opening edge of the jamb. Adjust the router's depth of cut to the combined thickness of the template and the hinge leaf. Then turn on the router and, holding it firmly in both hands, cut the mortise, keeping the base plate flat on the jig and the template guide flush against the edges of the cutout *(right)*. Move the router in small clockwise circles until the bottom of the mortise is smooth. Reposition the jig and rout the second mortise the same way. If the hinges you are using are rectangular, square the corners of the mortises with a chisel; for hinges with radiused corners, the mortises can be left rounded.

ROUTING HINGE MORTISES ON A DOOR

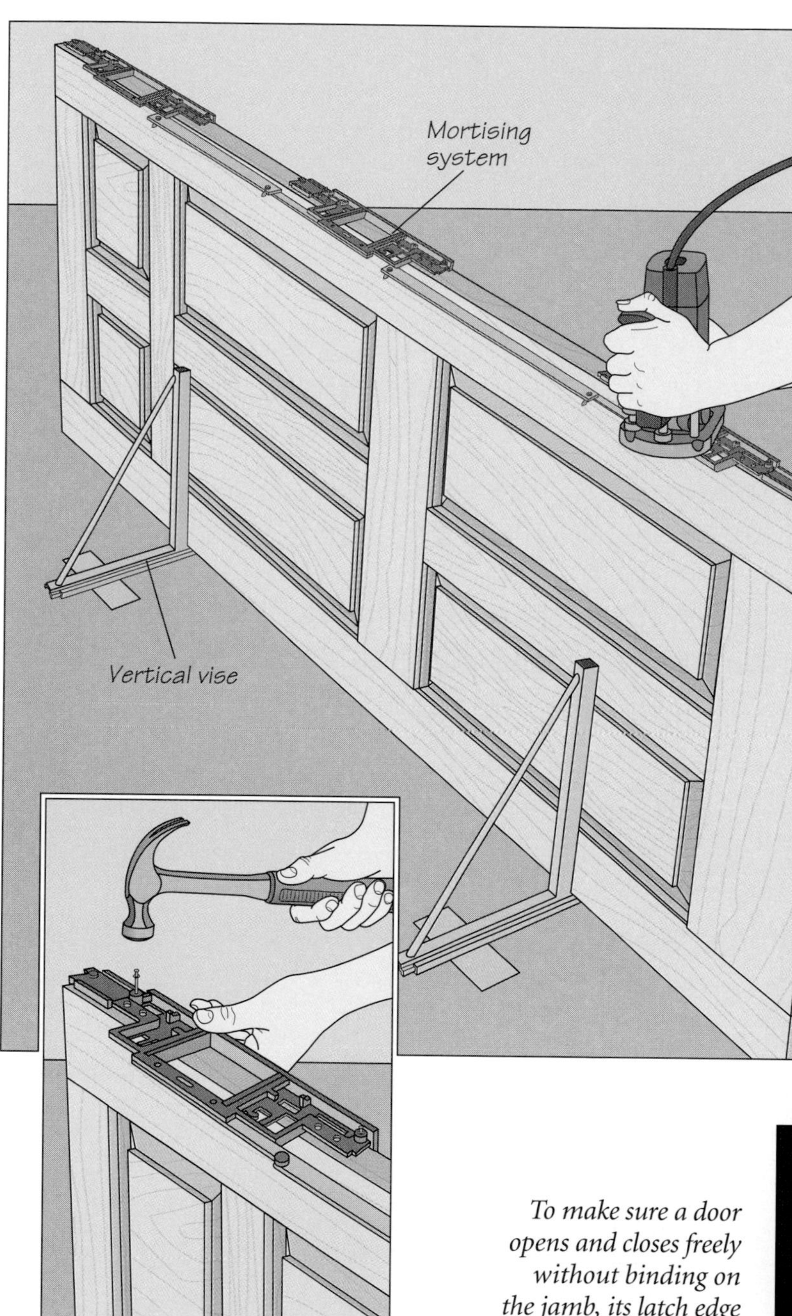

Routing the mortises

Hinge mortises on a door can be cut by hand *(page 100)*, or routed with the aid a shop-made jig *(page 101)* or a commercial mortising system, such as the one shown at left. The system can also be used to rout the mortises on the door jamb. Although the door shown will be hung with only two hinges, the jig features three adjustable mortise templates connected with metal rails. In this instance, the center hinge template is only being used to hold the jig together. Assemble the jig following the manufacturer's instructions, then secure the door edge-up with a pair of vertical vises or shop-made door bucks *(page 104)*. Mark the hinge mortises on the door edge and set up the jig, aligning the templates over your marks. Adjust the size of the templates for your hinges, making sure to compensate for the template guide you will use with your router. Tack the jig in place with the duplex nails provided *(inset)*. Install a ½-inch straight bit and a template guide in the router, turn it on, and cut each mortise by running the guide along the inside edges of the templates *(left)*. Square the corners of the mortises with a chisel.

To make sure a door opens and closes freely without binding on the jamb, its latch edge should be beveled slightly toward the side that contacts the doorstops. A portable planer, like the one shown in the photo at right, can be set to the desired bevel angle, enabling you to prepare several doors quickly.

MOUNTING A DOOR

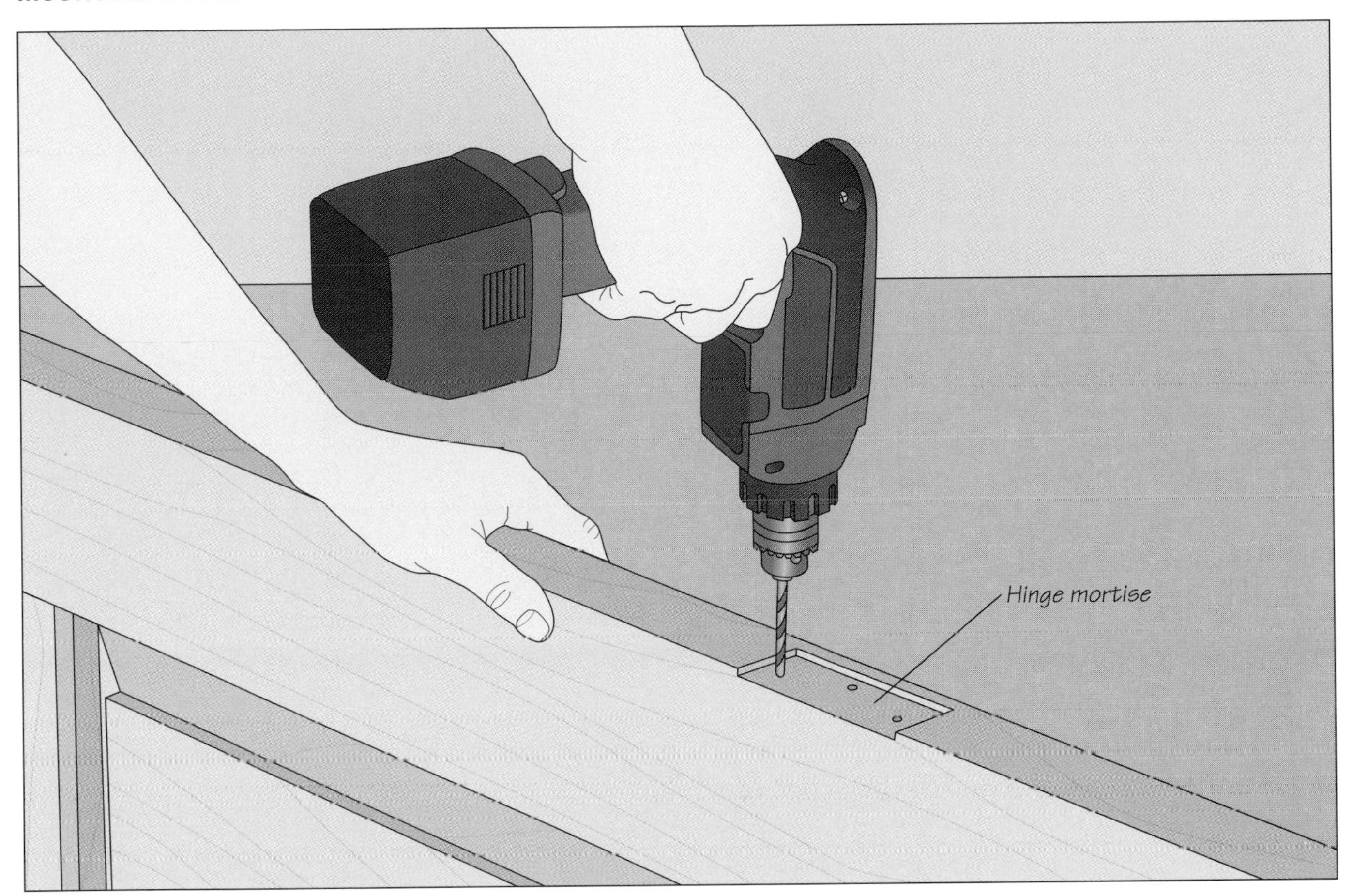

1 Drilling pilot holes for the hinge screws
Once all the mortises are cut in the jamb and the door, position each hinge leaf in place. Mark the screw holes with an awl, then bore a pilot hole at each location *(above)*, making sure you hold the drill as vertical as possible.

STANDARD HINGE SIZES

DOOR THICKNESS	Door width	Hinge height
1⅛" - 1⅜"	Up to 32"	3½" - 4"
	More than 32"	4" - 4½ "
1⅜" - 1⅞"	32" - 36"	5"
	36" - 48"	5" (Heavy-duty)
	More than 48"	6"
More than 1⅞"	Up to 43"	5" (Heavy-duty)
	More than 43"	6" (Heavy duty)

The chart above will help you choose hinges of appropriate height for your door. To determine correct hinge width (the combined width of the two leaves and the pin), first subtract the backset—the gap between the edge of the hinge and the door face—from the thickness of the door, then multiply the result by two. The typical backset is ¼ inch. The calculation for a 1¾-inch-thick door, for example, would be: (1¾" - ¼") x 2 = 3". Hinge sizes are expressed in height first, then width.

BUILD IT YOURSELF

DOOR BUCKS

Door bucks serve as an inexpensive alternative to commercial vertical vises for securing a door on edge for planing or cutting hinge mortises. The one shown in the illustration can be assembled quickly from plywood scraps. The dimensions provided will suit most doors.

To make the jig, start by cutting the jaws and feet from ¾-inch plywood, and the base from ¼-inch plywood. Screw the feet to the underside of the base flush with its ends. Then fix the jaws to the base, driving the screws from underneath. Countersink all your fasteners. Be sure the edges of the jaws align; leave a 1¾-inch space between them so the buck will hold any door of standard thickness.

Door bucks are usually used in pairs to secure a door edge up. Set the bucks on the floor a few feet apart, then slide the door between the jaws. The base will buckle slightly under the weight of the door, pulling the jaws together to grip the door and hold it in place *(right, bottom)*.

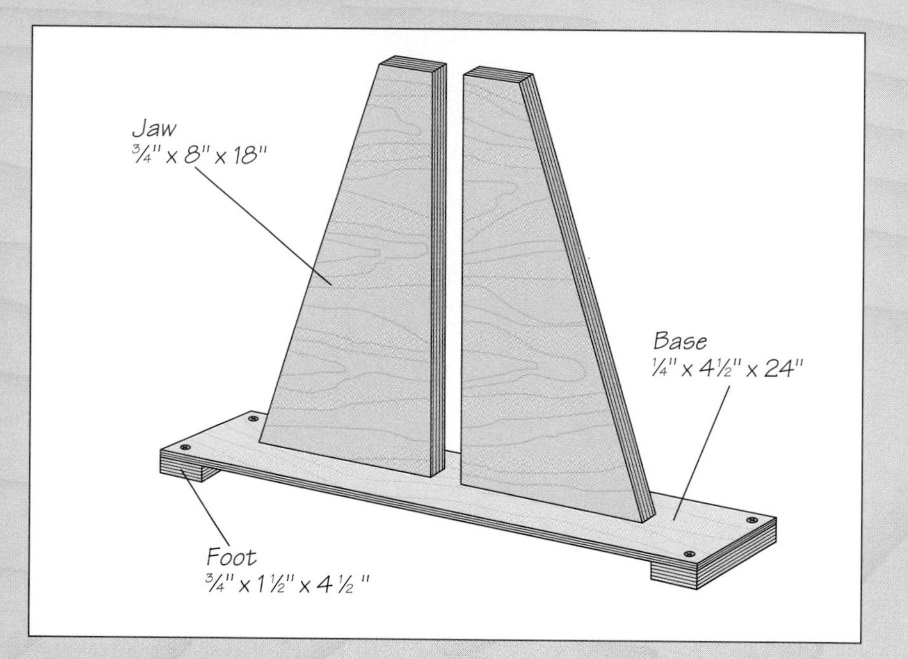

2 Mounting the hinge leaves
Remove the pins from the hinges and separate the leaves. Position one leaf in the mortise in the door edge and screw it in place *(right)*. Be sure to drive the screw heads flush with the hinge leaf. Fasten the other leaf to the jamb. Repeat at the remaining mortises.

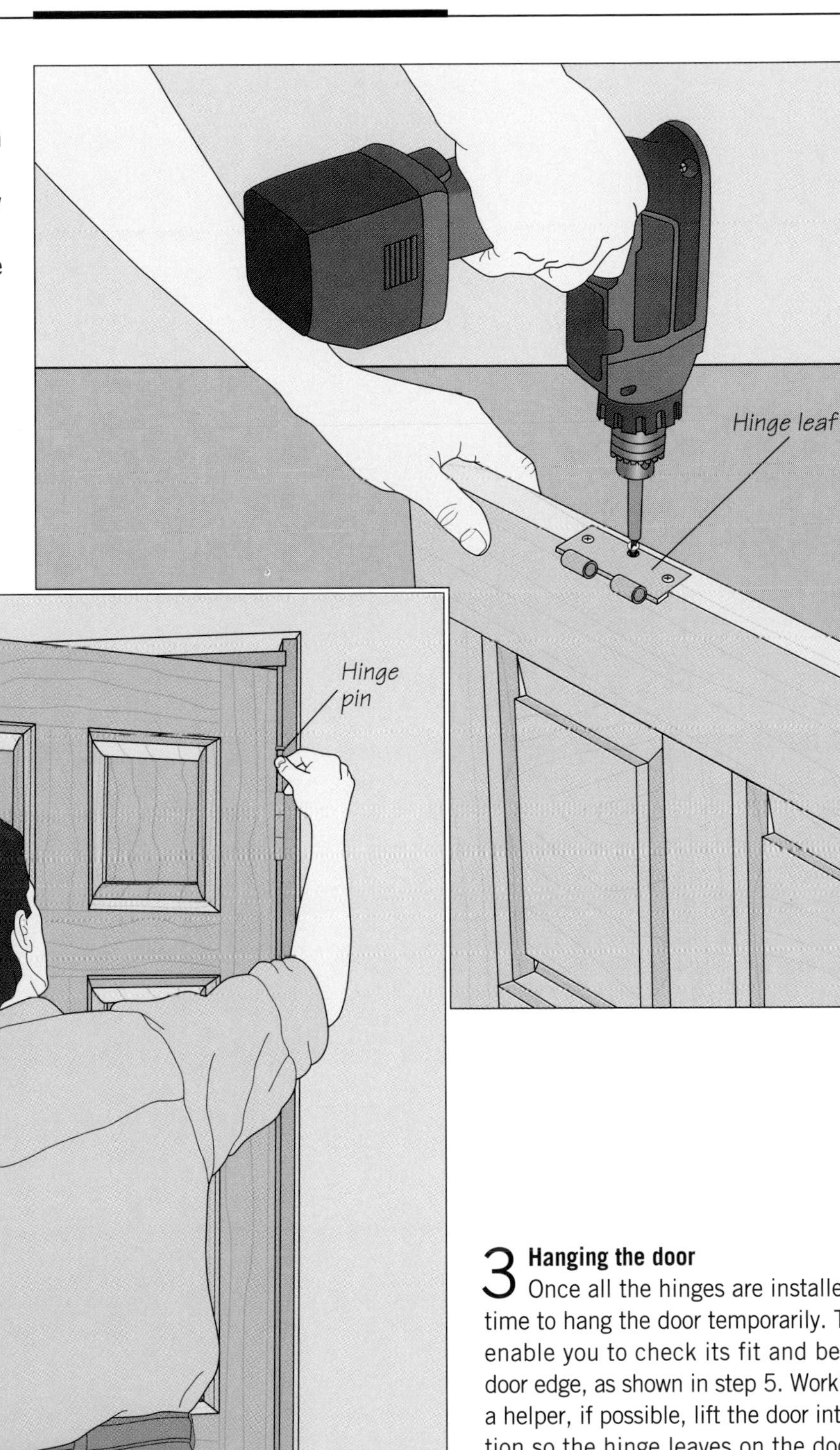

3 Hanging the door
Once all the hinges are installed, it is time to hang the door temporarily. This will enable you to check its fit and bevel the door edge, as shown in step 5. Working with a helper, if possible, lift the door into position so the hinge leaves on the door and jamb engage. If you have to work alone, hold the door upright and slide a few shims under its bottom edge. Bracing the door on the shims, join the top hinge leaves together. Then pivot the door to join the bottom hinge leaves. Slip each hinge pin partially in place *(left)* to lock the leaves together.

4 Marking the bevel
Doors typically require a 3° to 5° bevel on the latch edge to close properly. Stand on the doorstop side of the door and pull it shut. If the jamb was sized properly, the front edge of the door will hit the edge of the jamb, preventing the door from closing fully. To mark the bevel, hold the door against the jamb and use a pencil to scribe a line down the face of the door where it meets the side jamb *(left).*

5 Beveling the door edge
You can use a portable power planer to bevel the edge of the door *(page 102)*, but a jack plane will also work well. Remove the door from the opening and secure it latch-edge up. Transfer the bevel mark on the face of the door to the end. Then, starting at one end, guide the plane along the door edge *(right)*, walking next to the piece until you reach the other end. Hold the tool at the same angle as the marked bevel angle. Continue until you have cut to the line, then rehang the door. The door should contact the latch-jamb stop when you close it. Now install the lock-jamb doorstop *(page 99)*, butting it against the door and the head-jamb stop. Check the final fit of the door; there should be slight gaps between the door and the jamb. You can use the thickness of a dime and a nickel to measure these gaps. Pass a dime along the lock jamb to check for the 1/16-inch margin between the hinge jamb and the door. Use a nickel to measure the required 3/32-inch space at the top and along the other side. If necessary, remove the door again and plane down any spots where the gap is insufficient.

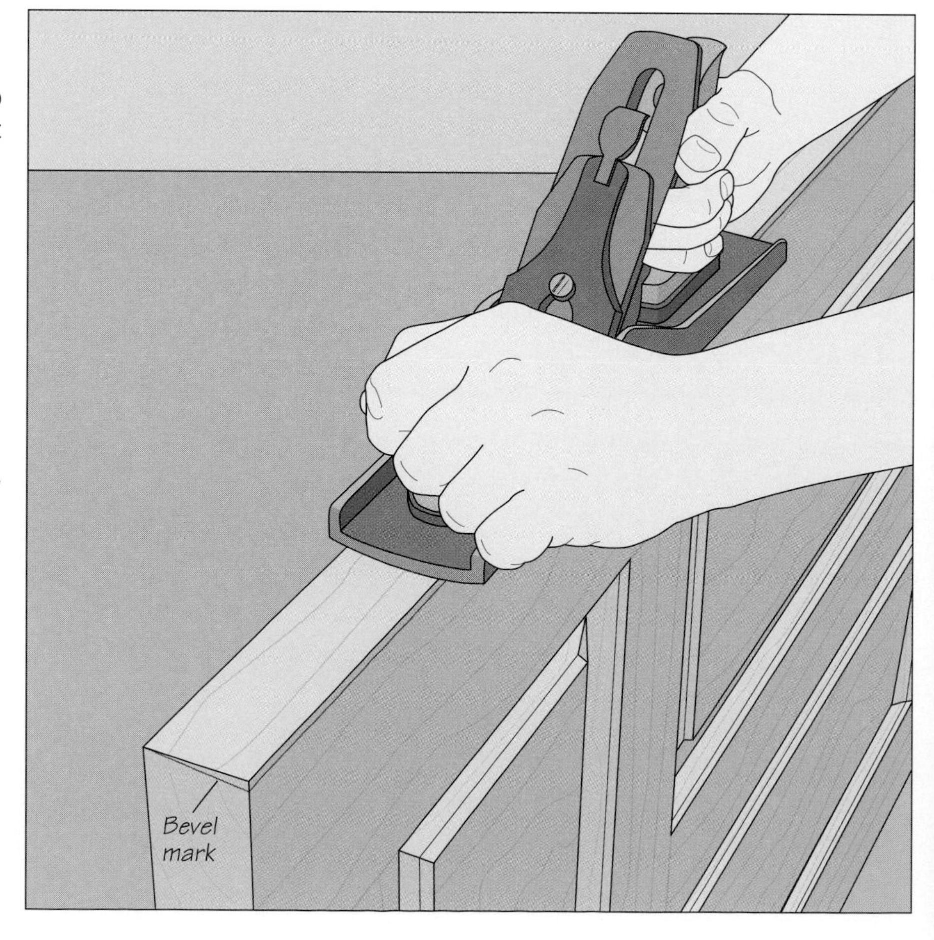

With the latch assembly fastened to the edge of the door, the doorknobs are fitted in place. The knob cover plate will then be screwed to the assembly to complete the lockset installation.

INSTALLING A PRIVACY LOCKSET

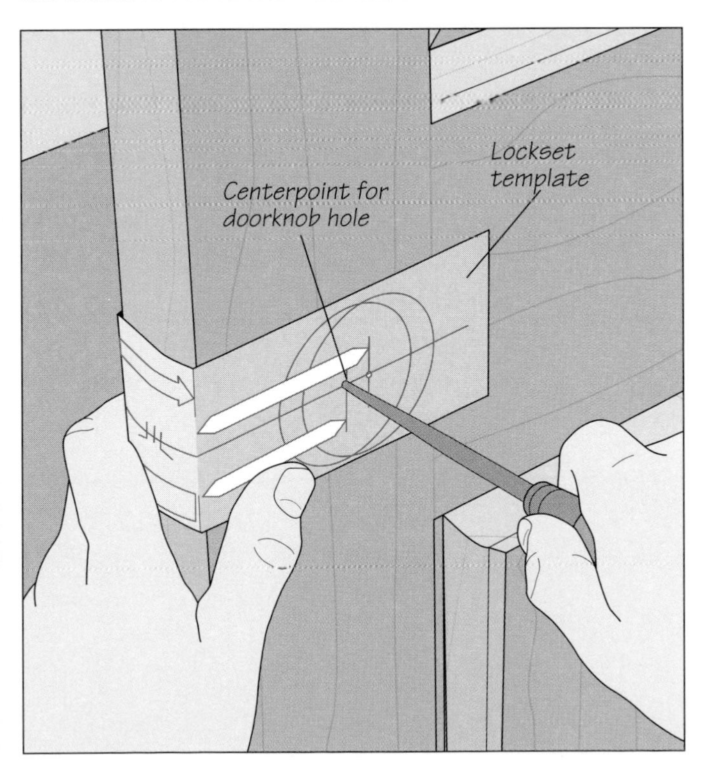

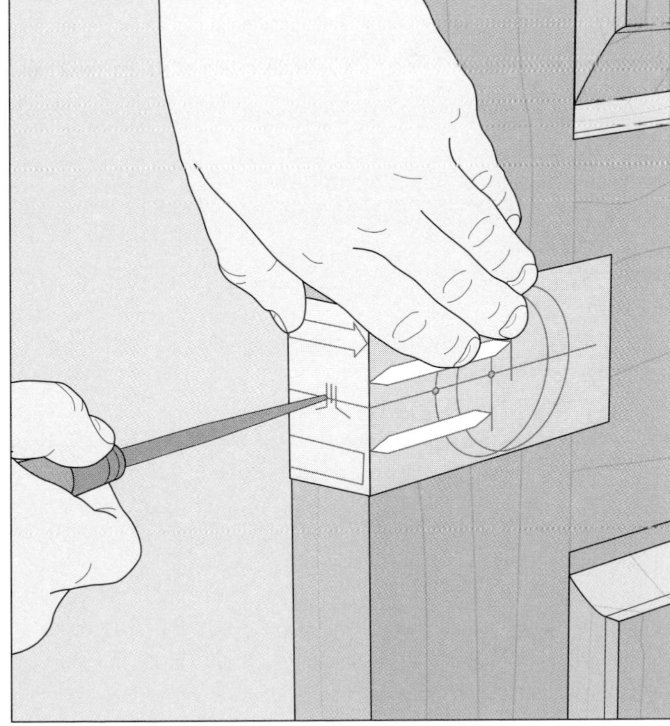

1 Positioning the lockset

Locksets usually come with a template for marking the holes you will need to drill for the latch assembly and doorknobs. Start by marking the height of the knobs on the door—typically 36 inches off the floor. Then tape the template over your mark. Use an awl to mark the doorknob point on the face of the door *(above, left)*—either 2⅜ or 2¾ inches from the door edge, depending on the model of lockset—then the centerpoint for the latch assembly hole on the door edge *(above, right)*.

2 Drilling the hole for the doorknobs
Install a hole saw in your electric drill, referring to the template for the correct diameter. The hole saw shown at right features a center pilot bit. Set the point of the pilot bit in the awl mark you made in step 1, then bore into the door until the pilot bit emerges from the other side. Keep the drill perpendicular to the door throughout. Now move to the other side of the door, insert the center pilot bit in the small opening you pierced through the door, and complete the hole. Drilling the hole in two steps will avoid splintering of the wood.

3 Boring the latch assembly hole
Replace the hole saw with a spade bit; again refer to the template for the appropriate bit diameter. Set the tip of the bit in the awl mark and bore the hole, keeping the drill perpendicular to the door edge *(below)*. For a narrow door, you can clamp wood blocks on the faces of the door on each side of the hole to prevent the wood from splitting. Stop drilling when you reach the handle hole. Some locksets require this hole to be drilled beyond the end of the doorknob hole for clearance.

4 Outlining the latch assembly faceplate
Slide the latch plate assembly into the hole you drilled in the edge of the door and set the faceplate flush against the door edge. Holding the faceplate square to the door edge, trace its outline with a pencil *(right)*.

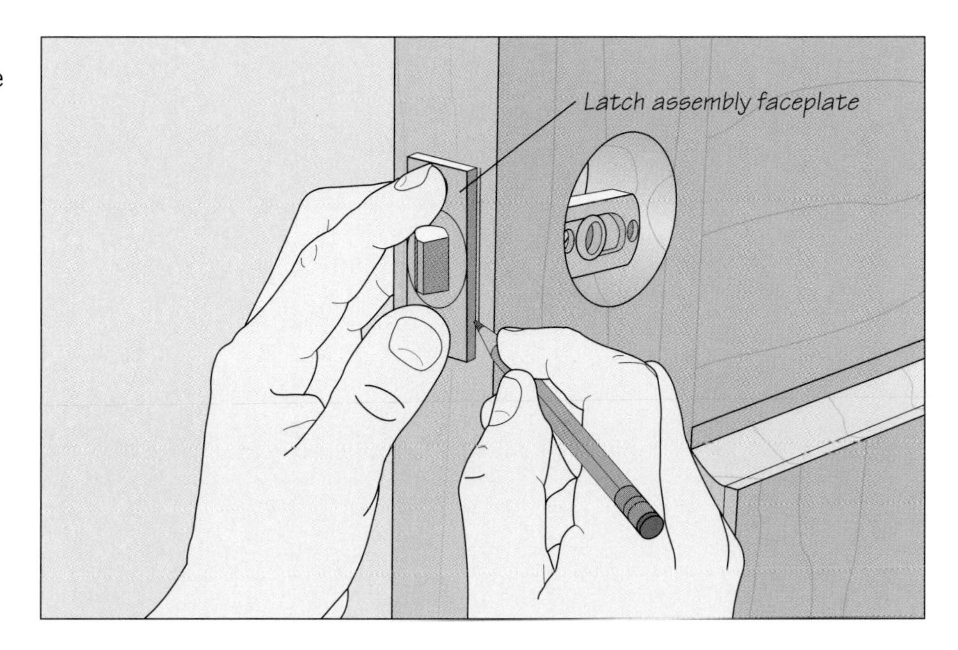

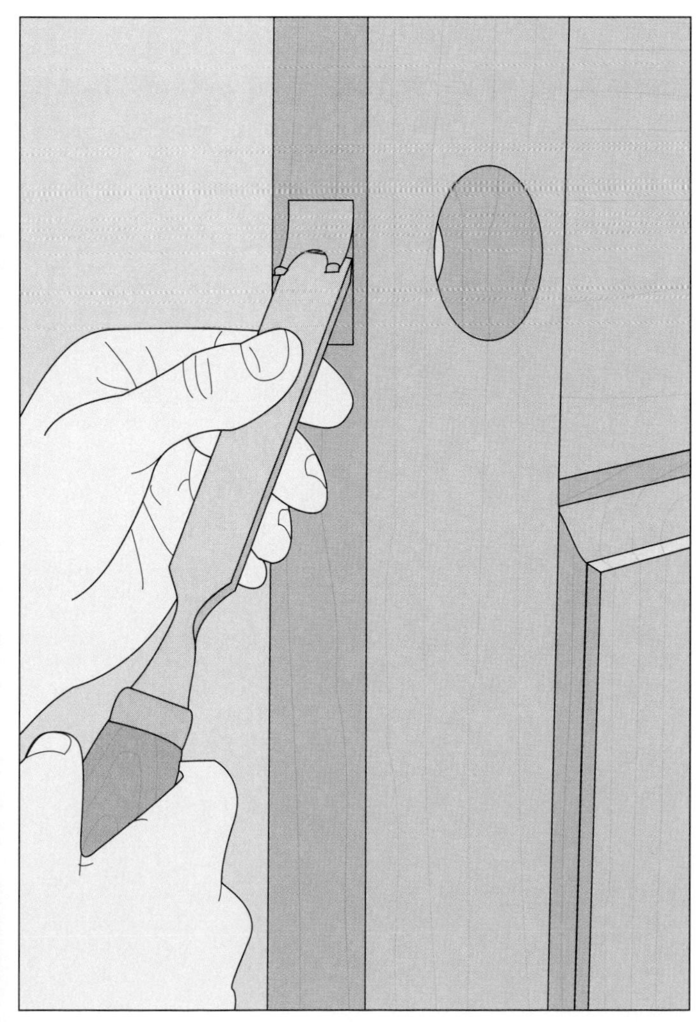

5 Installing the latch assembly
Use a chisel to cut a shallow mortise within the outline you marked in step 4. Start by scoring the outline of the mortise, then pare out the waste *(left)* to a depth equal to the thickness of the latch assembly faceplate. Using the chisel with the bevel facing down will help you control the depth of the mortise. Stop periodically and test-fit the faceplate in the mortise. Continue until the faceplate sits in the mortise flush with the door edge, then mark the screw holes with an awl. Drill a pilot hole at each mark. Finally, slide the latch assembly in the hole and screw the faceplate to the door edge *(above)*.

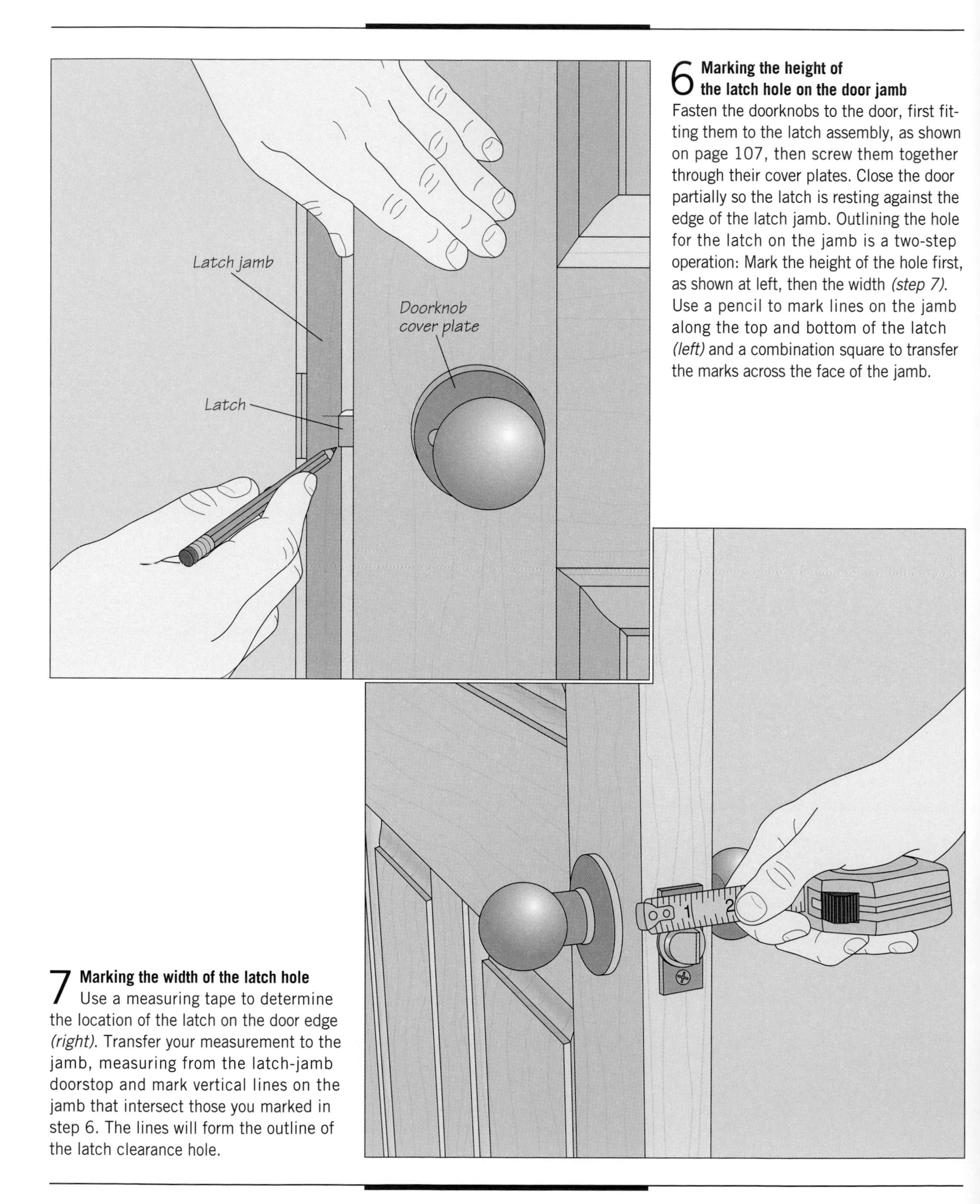

6 Marking the height of the latch hole on the door jamb
Fasten the doorknobs to the door, first fitting them to the latch assembly, as shown on page 107, then screw them together through their cover plates. Close the door partially so the latch is resting against the edge of the latch jamb. Outlining the hole for the latch on the jamb is a two-step operation: Mark the height of the hole first, as shown at left, then the width *(step 7)*. Use a pencil to mark lines on the jamb along the top and bottom of the latch *(left)* and a combination square to transfer the marks across the face of the jamb.

7 Marking the width of the latch hole
Use a measuring tape to determine the location of the latch on the door edge *(right)*. Transfer your measurement to the jamb, measuring from the latch-jamb doorstop and mark vertical lines on the jamb that intersect those you marked in step 6. The lines will form the outline of the latch clearance hole.

8 Cutting the latch clearance hole
You can use an electric drill fitted with a spade bit or, as shown at right, a chisel to form the latch clearance hole. The exact size of the hole is not critical, since the strike plate will cover most of it, but it must accommodate the latch when the door is closed. To use a chisel, first score the outline of the mortise, then clear out the waste *(right).* Test-fit the latch periodically by closing the door.

9 Installing the strike plate
Align the strike plate over the latch hole and mark its outline with a pencil *(left).* Then chisel a mortise within the outline to a depth equal to the strike plate thickness. Once the plate is flush with the jamb, hold it in position and mark the screw holes with an awl. Bore a pilot hole at each mark and screw the strike plate to the jamb. To check the installation, close the door. The face of the door should rest flush against the doorstops. If the door does not close properly, you can adjust the fit by bending the strike plate tongue slightly in or out.

INSTALLING BUTTED DOOR CASING

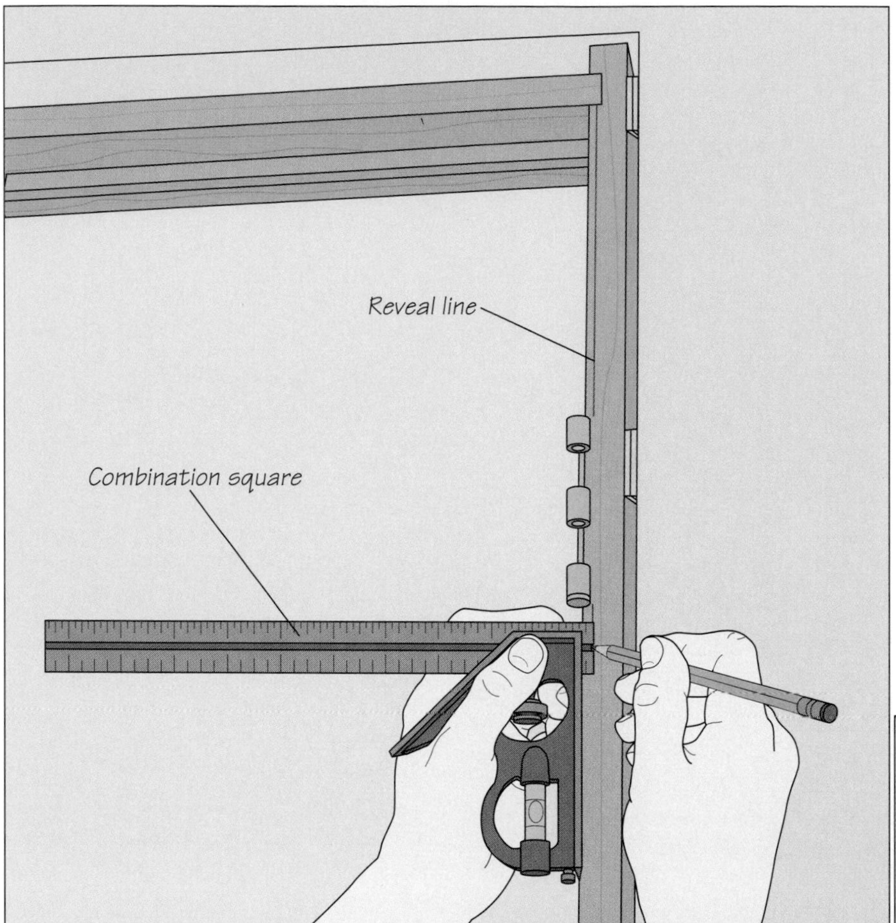

1 Marking the reveal
Before installing casing around a door, remove the door to give yourself enough room to work. Casing normally overlaps only a portion of the jamb edge, leaving part of it exposed. This exposed portion is known as the reveal. On ¾-inch-thick jambs like those shown, the reveal should be ⅛ to 3⁄16 inch. Make sure the casing will clear the hinges. To mark the reveal, use a shop-made reveal gauge *(page 63)* or a combination square. With the square, adjust it to the desired reveal width. Then, butting the square's handle against the top of the hinge jamb, hold a pencil against the end of the blade and run the square and the pencil down the jamb to the floor to mark the reveal *(left)*. Repeat the process on the other side jamb and across the head jamb.

2 Installing the plinths
Cut two plinths slightly thicker and wider than the casing you plan to use, and higher than the baseboard you plan to install on the wall. The plinth shown at right is cut from 1-inch-thick stock and beveled on one corner. Align the beveled edge of the plinth with the reveal line and fasten it to the wall and jamb with a hammer or a finish nailer gun. The plinth should rest flush on the finished floor. If the flooring has not yet been installed, set a piece of flooring under the plinth during installation to provide the required clearance. Repeat to install the plinth on the other jamb.

3 Installing the head casing
Cut the head casing so that it will extend slightly beyond both side-jamb casings. Align the casing with the reveal line on the head jamb and nail it to the wall and jamb *(right)*.

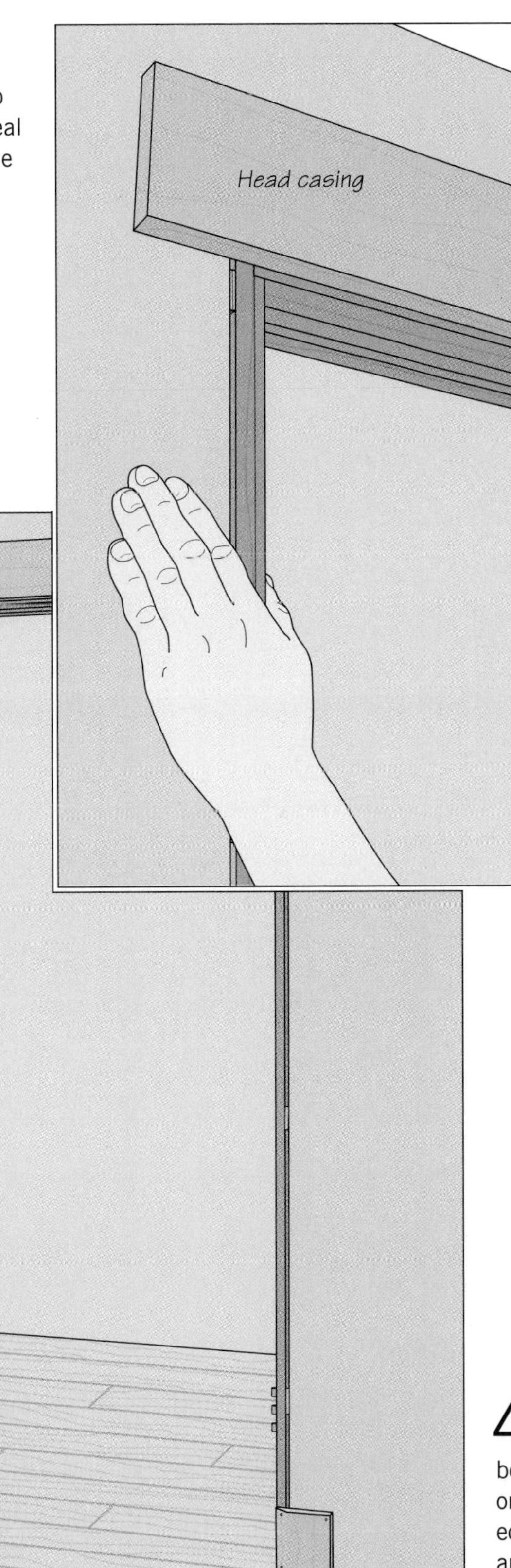

4 Installing the side casings
Cut the side casings to fit snugly between the plinths and head casing. Set one of the pieces in position, aligning its edge with the reveal line on the jamb *(left)*, and nail it in place. Repeat to install the second side casing. Now set all the nails.

STAIRS

A simple jig consisting of a 1-by-4 clamped to a carpenter's square is used to mark the rise-and-run — the width of the treads and risers—on a staircase stringer.

On a simple level, a run of stairs is nothing more than a conveyance from one floor to another. Functionally, a well-built staircase seems self-effacing—climbed or descended without thought or attention. But esthetically, a staircase may capture much attention. Precise joinery and ample use of fine woods help a staircase transcend rough carpentry and rise into the category of fine woodworking.

The elaborate stairways often associated with grand entrance halls, featuring curved handrails, goosenecked newel posts, and spiraling treads and risers belong to a bygone era. Craftsmen who specialize in this type of work are a rare breed. But even a simple straight-run staircase, like the one shown on page 116, can become the focal point of a home—and a challenging but feasible project for any woodworker. This chapter will show you how to get the job done, from design to installation.

Simple or complex, all staircases are built in much the same way: from the ground up. All you need is an opening in the floor above and a solid floor below. The staircase featured in this chapter also features a landing about halfway between floors, which should be made and incorporated into the wall framing before the stairs are designed.

The steps of the staircase—called the treads—rest on notched boards called stringers or carriages *(page 118)*. A staircase of typical width, about 36 inches, will usually need three stringers, one on each side with a third in the middle. Boards called risers close the vertical spaces between the treads *(page 123)*. The staircase is anchored at the top and bottom by newel posts *(page 128)*; the balusters *(page 136)*, or vertical posts between the steps and the handrail *(page 132)*, are doweled or dovetailed into the treads and nailed to the handrail.

Because poorly designed and executed stairs are accidents waiting to happen, building codes govern many aspects of their construction. For example, handrails should not be more than 34 inches above the treads. In some areas, wide stairs must have a handrail on both sides. Codes also carry stipulations governing headroom; 6 feet 8 inches is usually the minimum. Be sure to check with your local building code before embarking on a staircase.

A basic principle of staircase design is the so-called 17½ -inch rule. The combined width of one tread and height of one riser should equal 17½ inches. For the typical stair, this can mean a tread width of 10 inches and a riser height of 7 ½ inches, but depending on the needs of the stair's users, some variation is allowable.

As professional stairbuilder Scott Schuttner mentions in his introductory essay on page 10, consistency is important. All the treads in a staircase must be the same width and all the risers the same height. Even a slight deviation—while not easily discernable by the eye—will not go unnoticed by the person using the stairs.

A staircase can be as simple as a straight run of steps that lead into a cellar or as elaborate as the structure shown at left, with hardwood treads and risers, turned balusters, and a graceful curved handrail.

ANATOMY OF A STAIRCASE

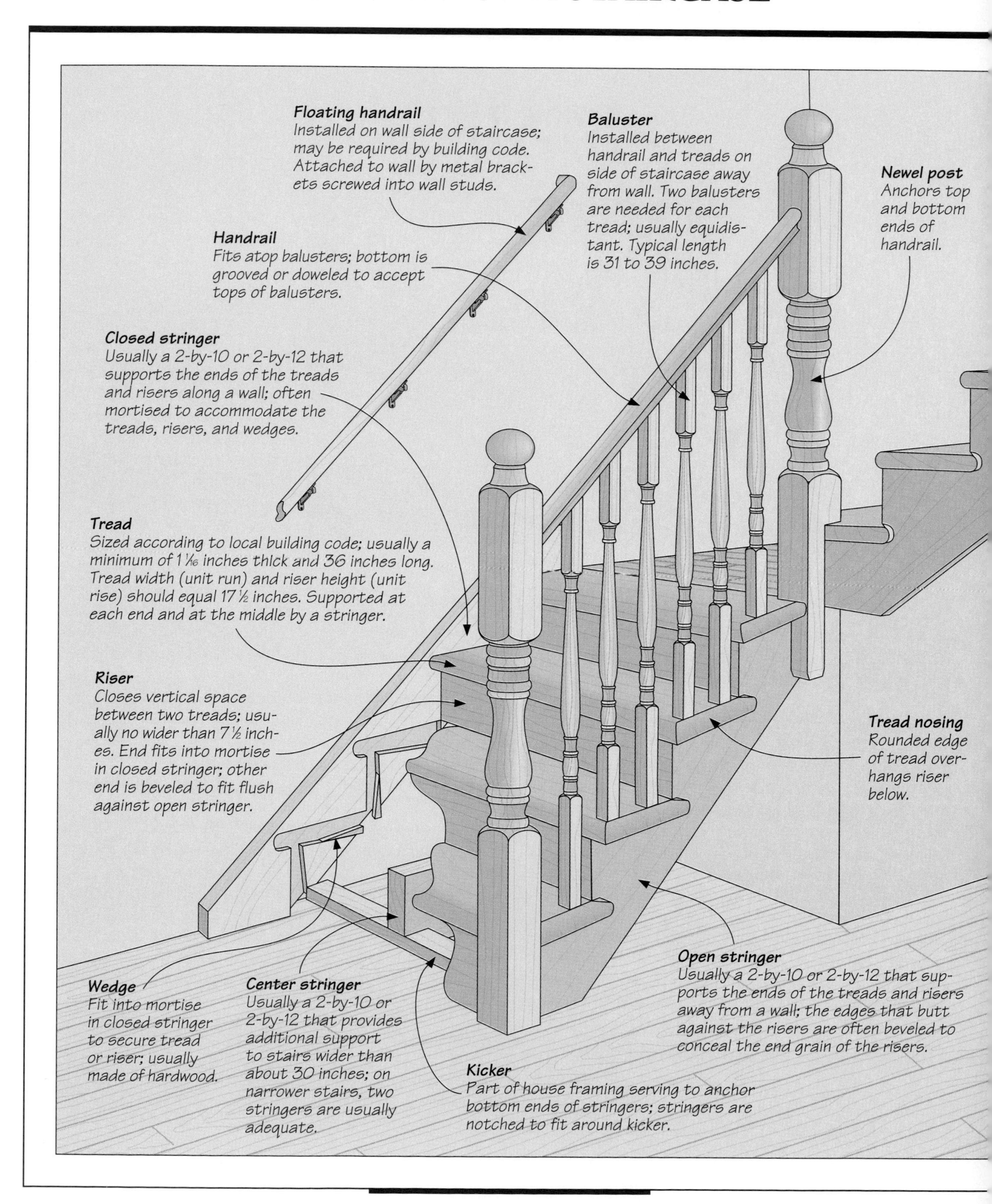

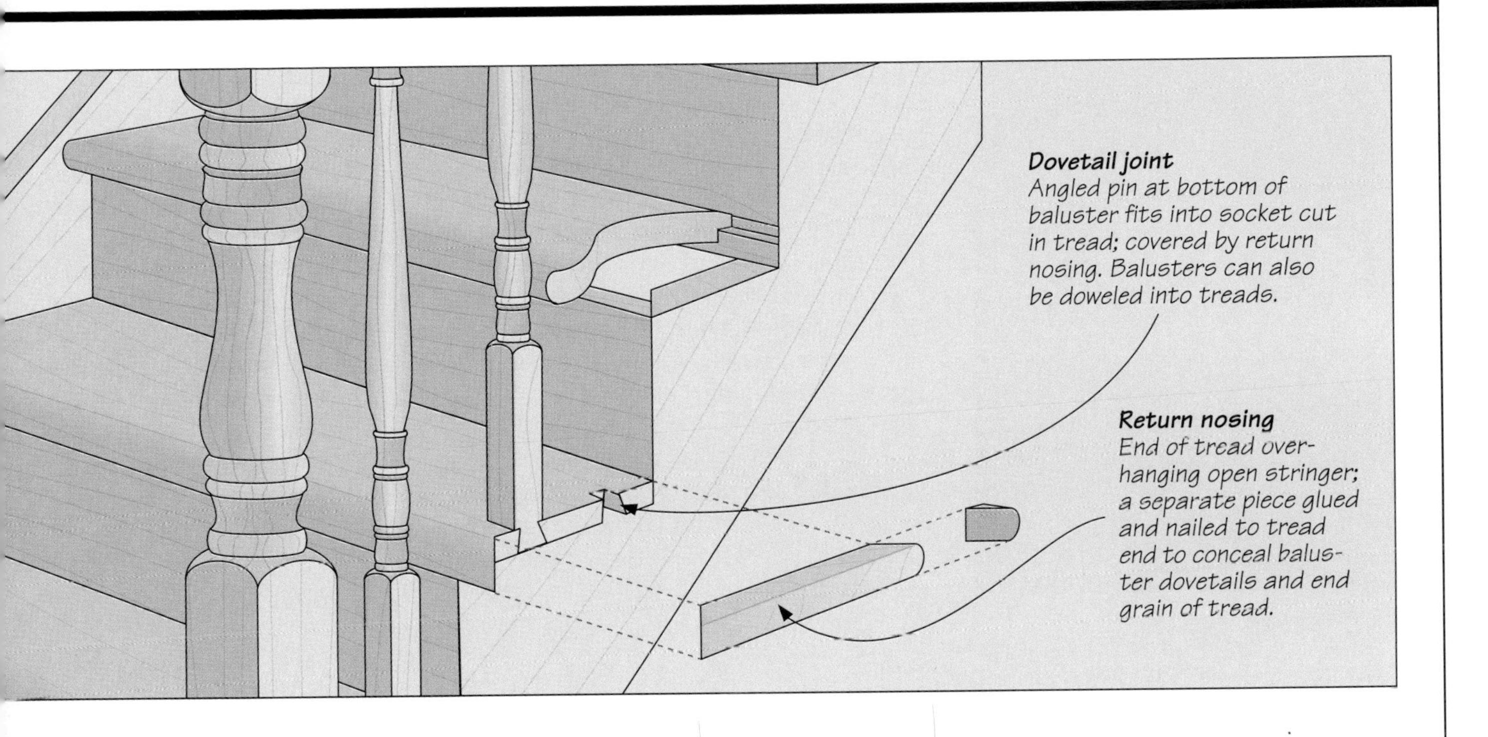

CALCULATING THE RISE-AND-RUN AND THE STRINGER LENGTH

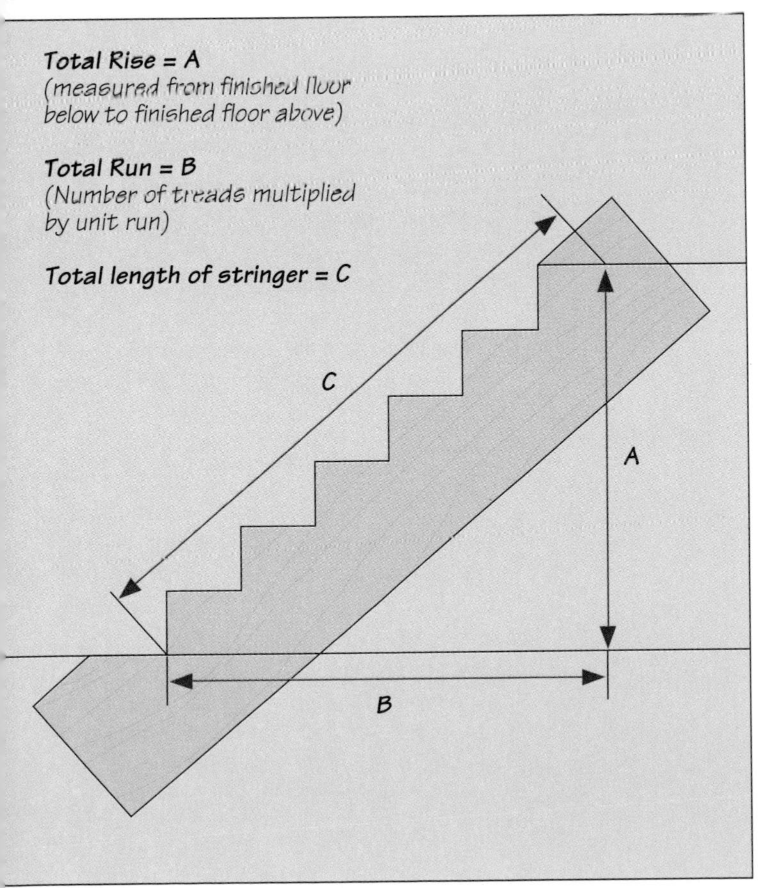

Determining the key dimensions

To calculate the length of the stringers, start by measuring the total rise—the distance from the finished floor below to the finished floor or landing above. Then divide your measurement by a whole number so that the result will be between 7 and 8 inches. If, for example, the total rise is 105 inches, dividing this measurement by 14 equals 7½ inches. The staircase would have 14 7½-inch-high risers. Next, use this result to determine the total run of the stairs, which is the total number of treads multiplied by the unit run. The staircase in our example would have the same number of treads as risers, 14. Since the 17½ inch rule dictates that the treads will be 10 inches wide (17½ less the riser height, or 7½), the total run equals 14 treads multiplied by 10 inches, or 140 inches. Once the you know the total rise and the total run, you can use the Pythagorean theorem and a pocket calculator to calculate the length of the stringers. The rise, run, and stringer of a staircase form a right-angled triangle with the total rise **(A)** and the total run **(B)** as the shorter sides and the stringer as the longest side, or hypotenuse **(C)**. The Pythagorean formula states that the squares of the shorter sides added together equals the square of the hypotenuse ($A^2+B^2=C^2$). In this case, square the total rise (105 x 105 = 11,025) and the total run (140 x 140 = 19,600) and combine the results (30,625). Then take the square root of this figure to arrive at the stringer length—in this example, 175 inches.

THE STRINGERS

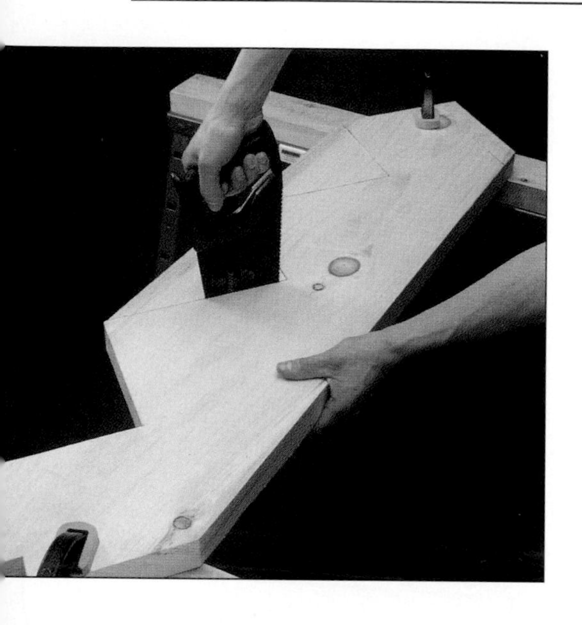

The stringers, also known as carriages, are the diagonal pieces that support the treads and risers. The principle behind sizing and notching them is a simple one that carpenters have known for centuries: For stairs to be ascended or descended comfortably and safely, an exact ratio must be maintained between the rise and the run—the distance users move up or down and the distance they move forward. This is often expressed as the "17 ½-inch rule": The sum of each rise-and-run should equal 17 ½ inches. (See calculations on page 117.)

Stringers can be either open or closed. An open stringer is simply a board with notches cut to support the treads and risers; it is usually used on the side of a staircase away from the wall. A closed stringer houses the ends of the treads and risers, often with mortises; the wall-side of a stairway usually has a closed stringer.

Most of the notching of the center stringer can be done with a circular saw, but use a crosscut saw to complete the job. Make sure you hold the saw vertically as you cut into the corners.

MAKING AND INSTALLING THE STRINGERS

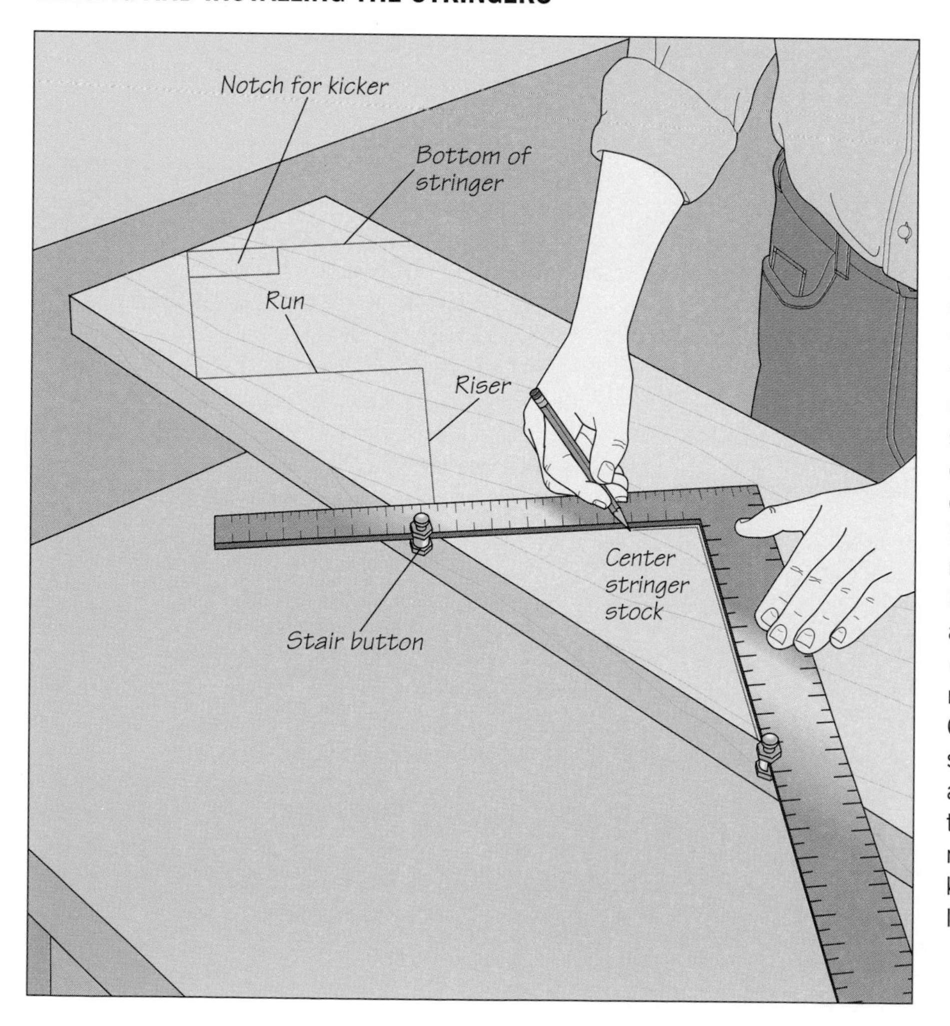

1 Marking the center stringer
Once you have determined the unit rise and unit run of your stairs and calculated the stringer length, prepare the stringers. Cut them to length from 2-by-12 stock. Start by laying out the center stringer; you will use it as a template to mark the others. Set one board face down on a work surface. To lay out the rise-and-run, attach two commercial stair buttons to a carpenter's square, positioning one for the rise and the other for the run. If you do not have stair buttons, use the shop-made jig shown on page 115. Then, starting about 12 inches from one end of the board, hold the stair buttons against one edge of the stock and mark the first unit rise-and-run along the inside edge of the arms of the square with a pencil. Slide the square along and repeat *(left)*, ensuring that the next unit rise-and-run starts exactly where the first one ends. Continue marking until you reach the opposite end of the board. Once all the steps are marked, add cutting lines at 90° for the top and bottom of the stringer. Also mark the notches that will fit around the kicker at the staircase bottom and the ledger board at the top.

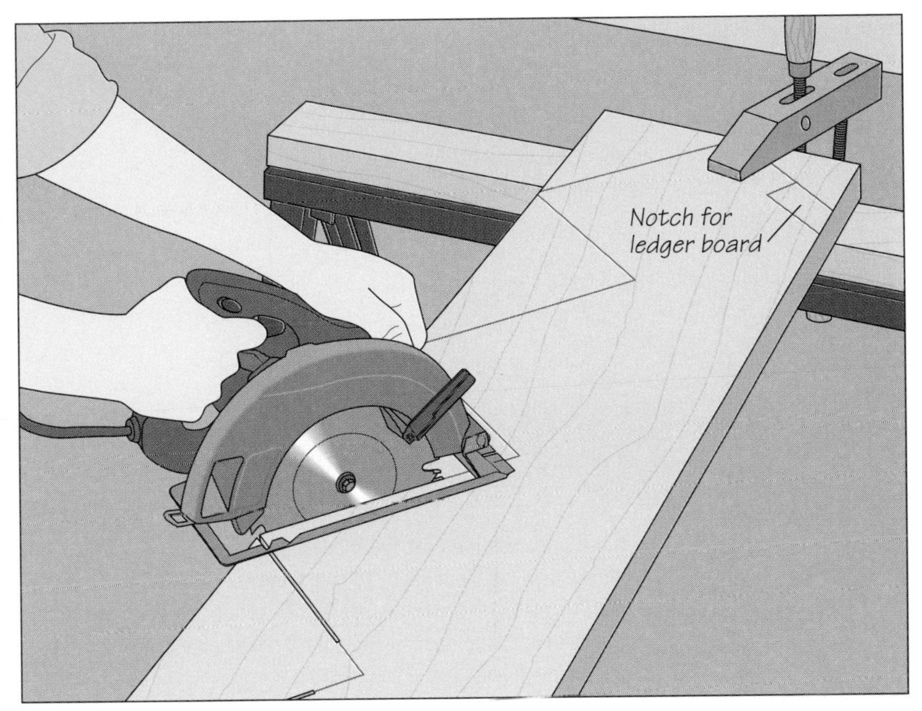

2 Cutting the center stringer
Clamp the stringer board face up across a pair of sawhorses and cut along your marked lines using a circular saw *(left)*. When you reach the ends of the board, reposition it on the sawhorses as necessary. Do not try to cut right to the corners of cutting lines with the circular saw. Instead, finish the cuts with a handsaw *(photo, page 118)*.

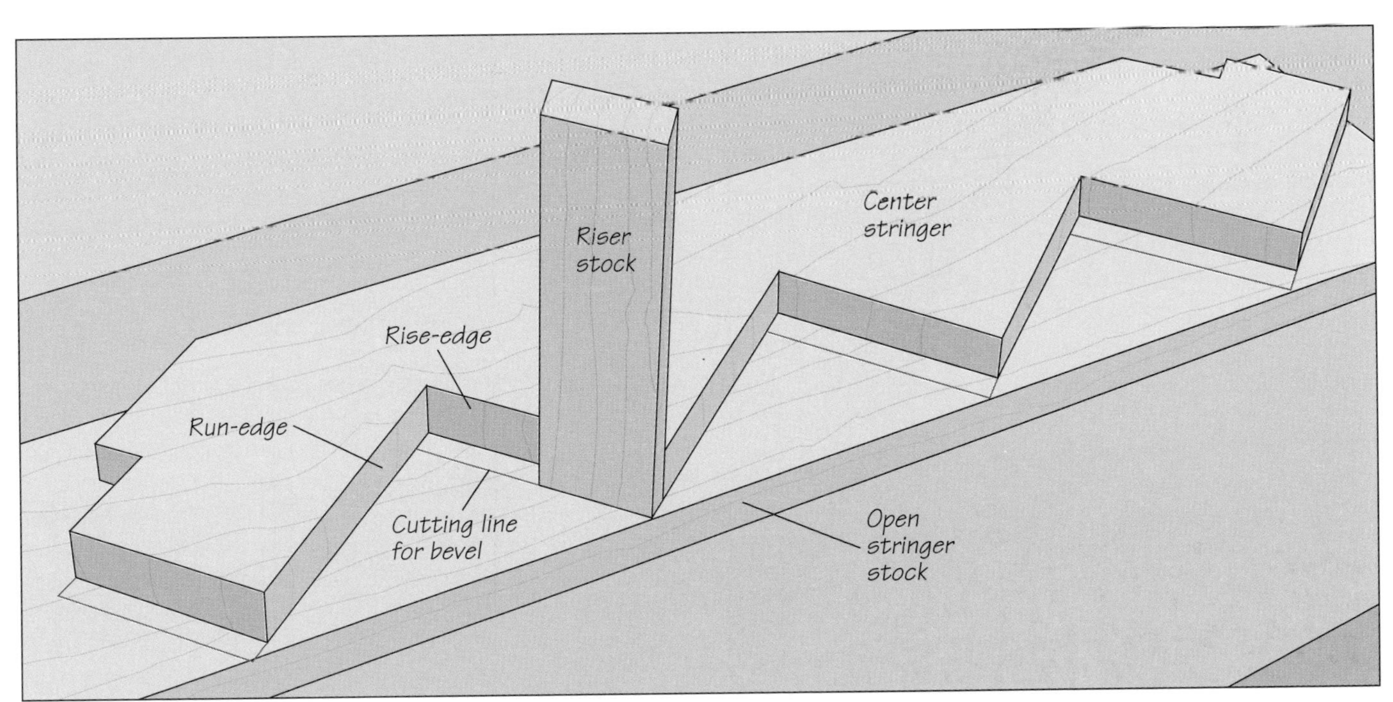

3 Laying out the open stringer
Set the open stringer board on a work surface and lay the cut-out center stringer on it, leaving a gap between the edges of the two boards equal to the width of your riser stock. Mark the run edges of the center stringer on the open stringer board. To mark the rise edges of the stringer, take a piece of riser stock and set it on end on the open stringer board, butting its face against one of the rise-edges of the center stringer. Mark a line along the riser board from the edge of the open stringer board to the run-edge of the center stringer *(above)*. Repeat at all the other rise-edges of the center stringer. This second set of cutting lines will compensate for the bevel you will need to saw in the rise-edges of the open stringer.

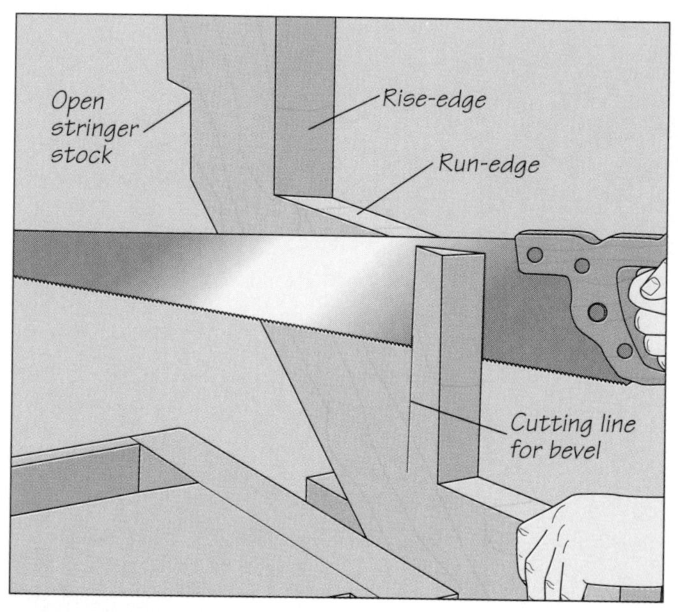

4 Cutting the open stringer

Cut out the open stringer the way you cut the center stringer, following the cutting lines you marked in step 3. For the bevel cuts, secure the stringer stock end up in a bench vise. At the outside corner of each step, use a combination square and a pencil to mark a line at a 45° angle across the run-edge of the stringer. Then extend the line at a 90° angle down the inside face of the stringer. Use a crosscut saw to cut the bevel (above), stopping when you reach the bottom of the rise edge. Remove the waste with a horizontal cut. Bevel the other rise-edges of the open stringer the same way.

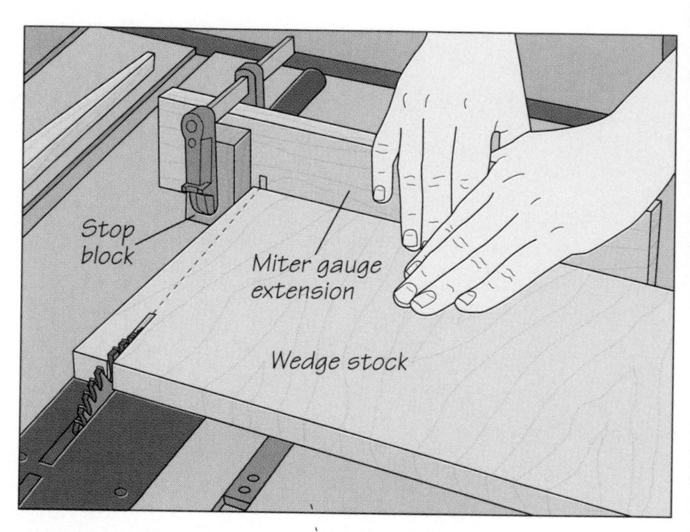

5 Making the wedges for the closed stringer

Before you can outline the mortises in the closed stringer, you have to make the wedges that will support the treads and risers in the mortises. Use ¾-inch-thick hardwood for the wedges and cut them on your table saw. Screw a board to the miter gauge as an extension and clamp a stop block to the extension about ⅛ inch from the blade. Angle the miter gauge so that you will cut wedges that will taper from about ⅞ inch thick to about ⅛ inch thick. Holding your wedge stock flush against the miter gauge extension so that its grain is parallel to the blade, cut the wedges *(above)*. Turn the board over after each pass. You may need to make a few test cuts and adjust the miter gauge angle until the wedges are the right size. **(Caution: Blade guard removed for clarity.)**

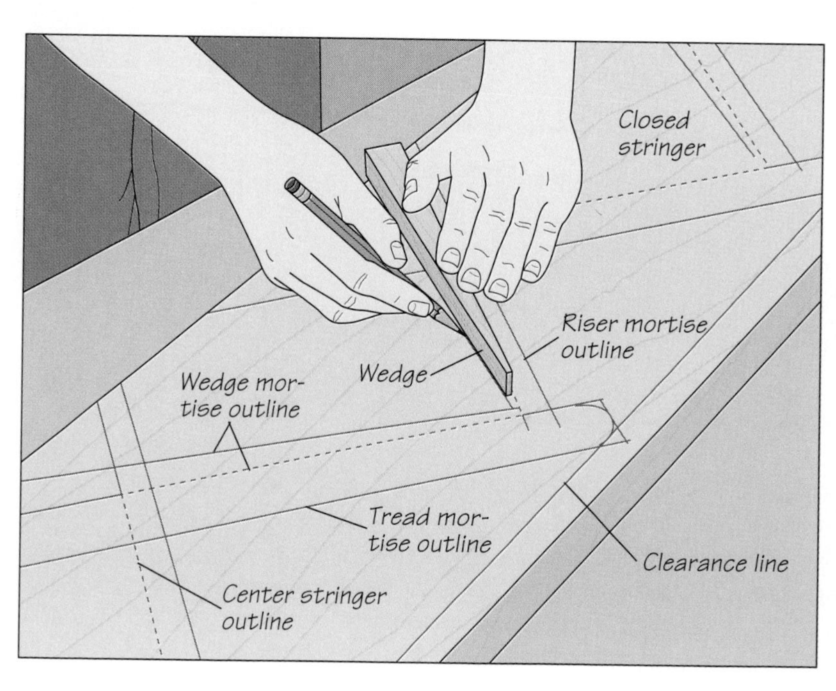

6 Marking the mortises on the closed stringer

The mortises in the closed stringer must take into account the treads, the risers, and the wedges. Since the treads are housed in the closed stringer, start by marking a clearance line along your stringer stock about 2 inches below the top edge. Then, outline the center stringer on the closed stringer board, allowing enough space so that when the tread mortise is cut it will not project beyond the clearance line. Position a piece of tread stock on end on the board flush with the center stringer outline and mark its outline at each tread location. Repeat with a piece of riser stock. Finally, position a wedge flush with each tread and riser outline and mark its outline *(left)*. Make sure to position the thin end of the wedge at the joint between the tread and riser. Extend all your lines to the bottom edge of the stringer.

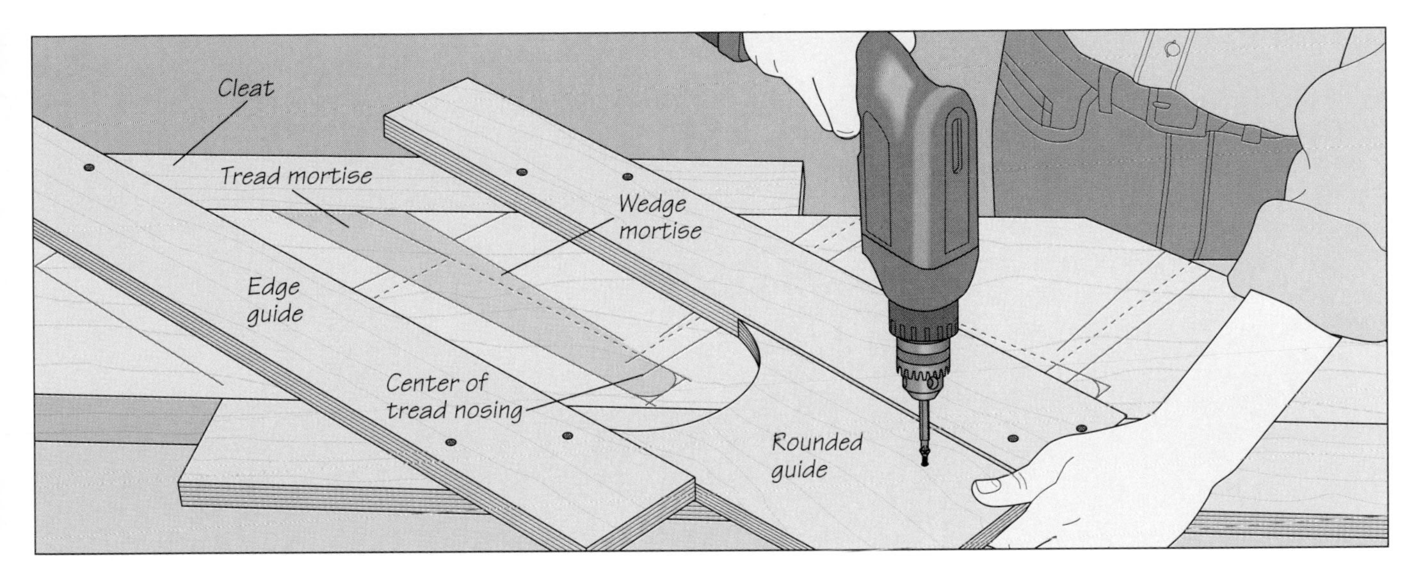

7 Making the closed stringer mortising jigs

It is easiest to rout the mortises in the closed stringer with the help of a jig. Although commercial jigs are available for this task, you can easily build your own. Make one jig for the mortises housing the treads and the wedges under them, and a separate jig for the risers and the wedges behind them. Make the tread-mortising jig first. Start by securing the stringer face up on a work surface and positioning cleats along its edges. Install a ½-inch straight bit in your router and align the bit with one edge of a tread outline. Screw a board as an edge guide to the cleats parallel to the treads so that its edge is flush against the router base plate. Repeat with a second edge guide parallel to the opposite edge of the tread outline. Since the front edges, or nosings, of the treads are rounded, you will need a rounded edge guide along the front of the tread mortise. Adjust a compass to the distance between the edge of the router bit and base plate. Then, holding the compass point at the center of the tread nosing, draw an arc on the stringer. Cut the third guide to fit between the edge guides, sawing an arc equal to the tread nosing and the router base plate out of one end, then screw it in place *(above)*.

8 Routing the mortises in the closed stringer

Align the tread-mortising jig over one of the tread-and-wedge outlines and clamp it in place. Plunge the router bit into the stock, riding the base plate against the edge guides to rout the edges of the mortise. Move the router in small clockwise circles to remove the remaining waste from the mortise *(above)*, stopping when the bottom of the cavity is smooth. Repeat for the remaining tread-and-wedge mortises. Make the riser-mortising jig the same way, omitting the rounded edge guide, since the front edges of the risers are square. Then rout the riser-and-wedge mortises.

9 Installing the closed stringer
Once all the mortises in the closed stringer have been routed, you can install the stringers. Start with the closed stringer. Hold it flush against the wall and fasten it to the wall, driving screws into every wall stud. Use screws that are long enough to reach the studs, making sure to drive the fasteners below the tread mortises *(right)*. This way, the screw heads will not be visible.

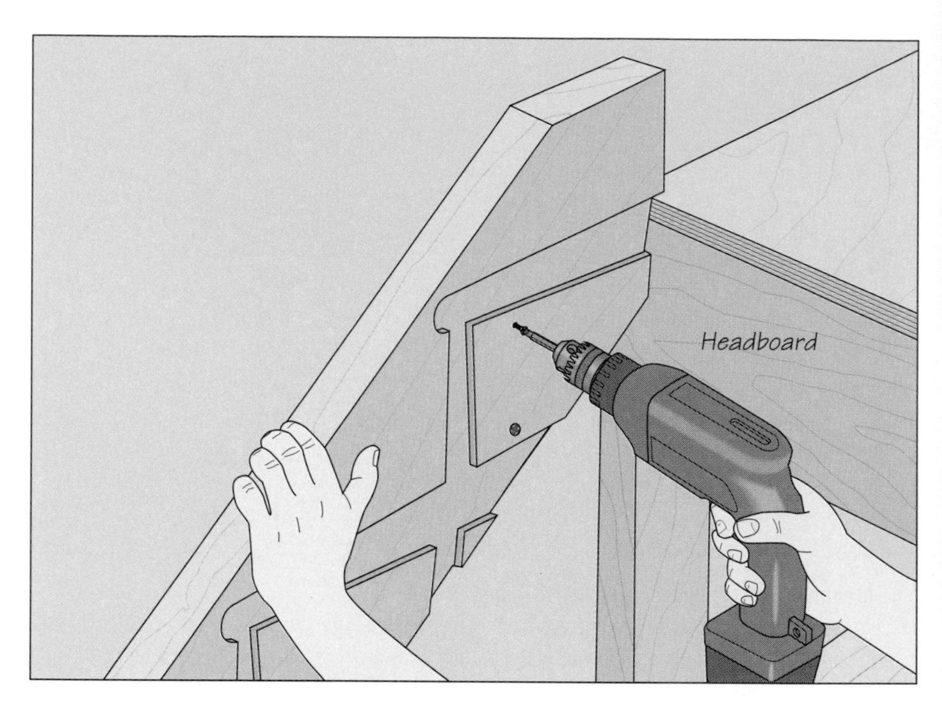

10 Installing the open and center stringers
Butt the open stringer against the headboard and mark a line on the headboard along the top edge of the notch in the stringer. Cut a 2-by-4 to the width of your staircase. Align the board with the marked line on the headboard and use lag screws to attach it to the header. Check with a carpenter's level to make sure the board is level as you fasten it in place. The 2-by-4 will serve as the ledger board to support the top of the open and center stringers. Repeat to position and install the kicker at the bottom of the stringers. Once the ledger board and kicker are in place, set the open stringer in position and fasten it to the two boards, driving screws at an angle through the stringer. Repeat to install the center stringer *(left)*.

TREADS AND RISERS

Anatomy of a tread and riser
Although the treads and risers in prefabricated wood stairs are often simply butted together and joined with screws, classic stairbuilding uses sturdier joinery techniques, as shown at right. Tread stock is available pre-milled; all you need to do is cut to length. If you are making treads from rough lumber, cut them 1½ inches thick if you are using softwoods and 1¼ inches thick for hardwood. To prepare the treads for assembly, cut a rabbet along the back edge of each one *(page 124)*; the resulting lip will fit into a groove in the riser above. Then rout a groove in the tread's underside near the front edge to accommodate the riser below. For the risers, cut the groove in the front face near the bottom to accommodate the lip in the tread below. Bevel the end that will fit against the bevel on the open stringer *(page 120)*. Once the treads and risers are assembled and glued together, cut triangular glue blocks to reinforce the joints. Glue and nail two or three blocks at the back of each tread-riser joint, locating them near the stringers. Finally, make a length of cove molding for each tread-riser joint and glue and nail it to the front of the joint. The cove molding is strictly decorative.

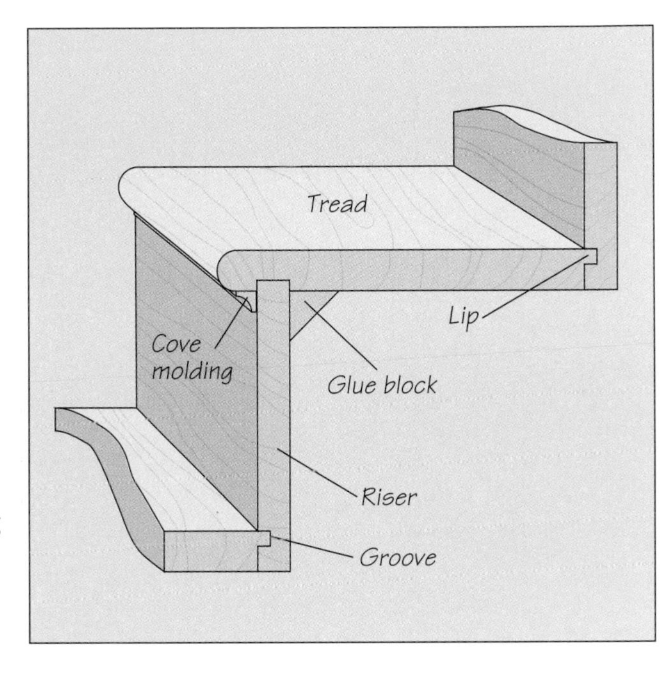

PREPARING THE TREADS AND RISERS

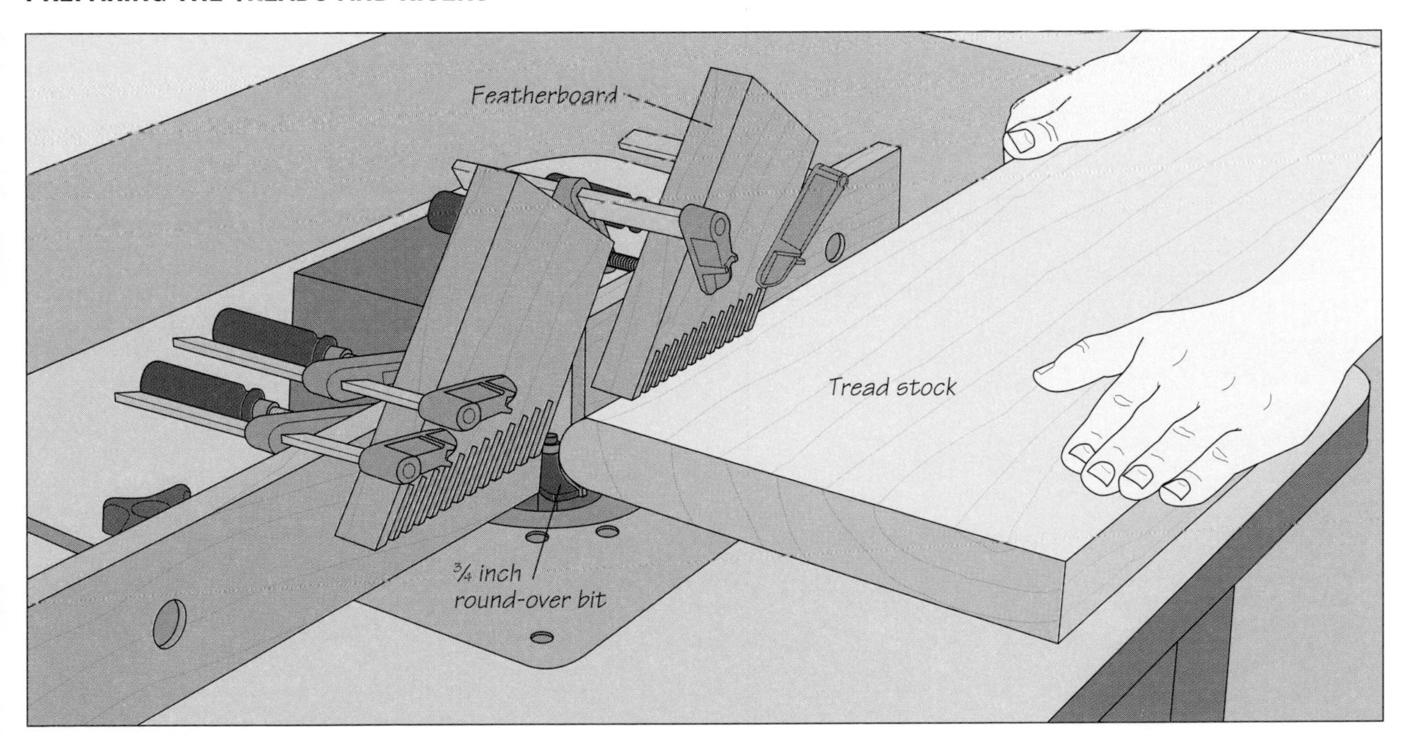

1 Rounding over the front edges of the treads
Once your treads are cut to length, you need to shape the front edge of each one to form the nosing that overhangs the riser below. Install a ¾-inch piloted round-over bit in your router and mount the tool in a table. Align the fence with the bit's pilot bearing and clamp two featherboards to the fence, one on each side of the cutter. Starting with a shallow cutting depth, feed the tread across the table, pressing the front edge against the fence and the pilot bearing. Make several passes on each face, increasing the depth of cut by ¼ inch each time until the edges are rounded over *(above)*.

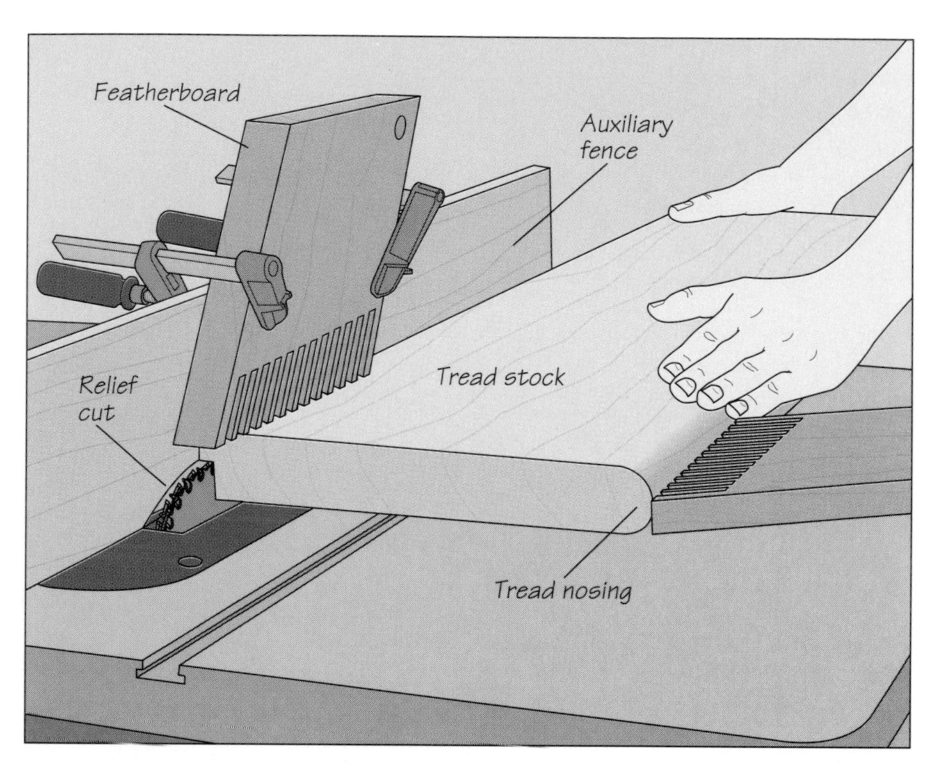

2 Cutting the rabbets at the back of the treads

Make the joinery cuts in the treads on your table saw. Install a dado blade, adjusting its width to slightly more than ¼ inch. Set the depth of cut so the resulting lip in the tread will fit snugly in the groove in the riser *(see anatomy, page 123)*. Then attach an auxiliary fence to the saw and cut a relief notch in it. Use two featherboards to brace the treads, clamping one to the fence above the dado head and another to the table in line with the blades. Position the fence for a cutting width of ¼ inch and feed the tread across the table, holding the back edge flush against the fence *(left)*. **(Caution: Blade guard removed for clarity).**

3 Cutting the grooves in the treads

Once all the rabbets are cut in the treads, readjust the dado head to a width of ¾ inch—the thickness of the riser stock. Set the cutting depth at ¼ inch and reposition the fence to locate the groove 1¼ inch from the tread nosing. Again, use two featherboards to brace the treads, repositioning the one on the saw table as necessary. Feed the tread into the dado head, keeping the nosing flush against the fence *(above)*. **(Caution: Blade guard removed for clarity).**

4 Preparing the treads for return nosing
To hide the end grain of the treads at the open-stringer end, cut a piece of stock from the end of each tread, leaving a bevel that will mate with the return nosing *(page 139)* that is applied once the stairs are assembled and the balusters are installed. Clamp a tread astride sawhorses as shown. To outline your cut, mark a straight line on the top of the tread 1¼ inches from the open-stringer end. Then mark another line at a 45° angle starting from the front corner and intersecting the first line. Use a crosscut saw to make the cuts, starting with the bevel *(right)*. Repeat for the other treads.

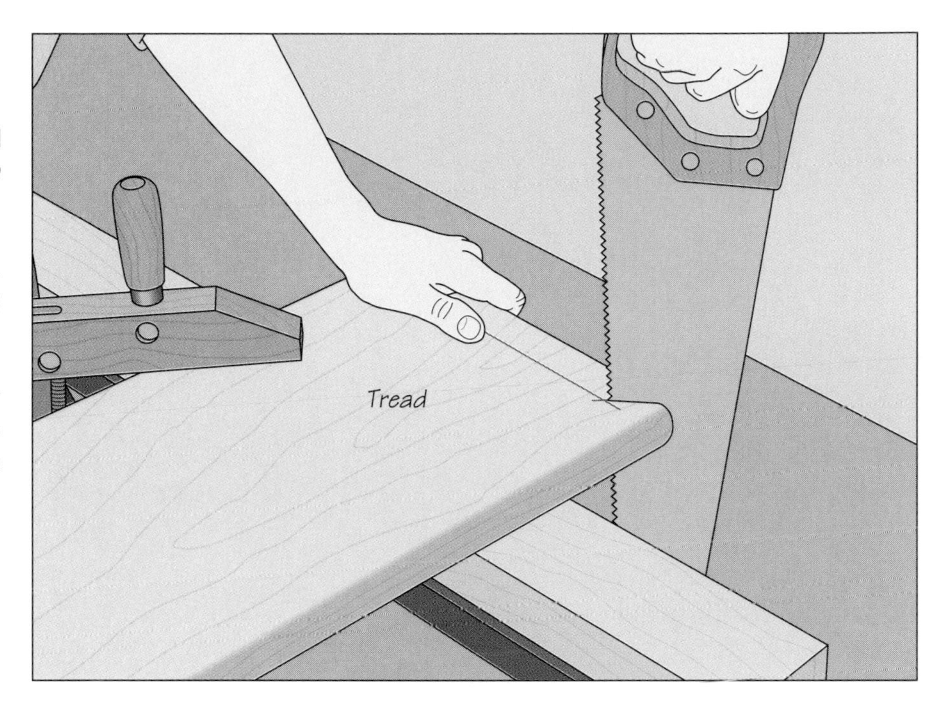

INSTALLING THE TREADS AND RISERS

1 Installing the bottom riser
Prepare a riser with no groove in its front face and set it in position with the beveled end flush against the open stringer and the straight end seated in the mortise in the closed stringer. Use finishing nails to fasten the riser to the open and center stringers. Drive a wedge into the closed stringer mortise behind the riser to secure it in place *(left)*.

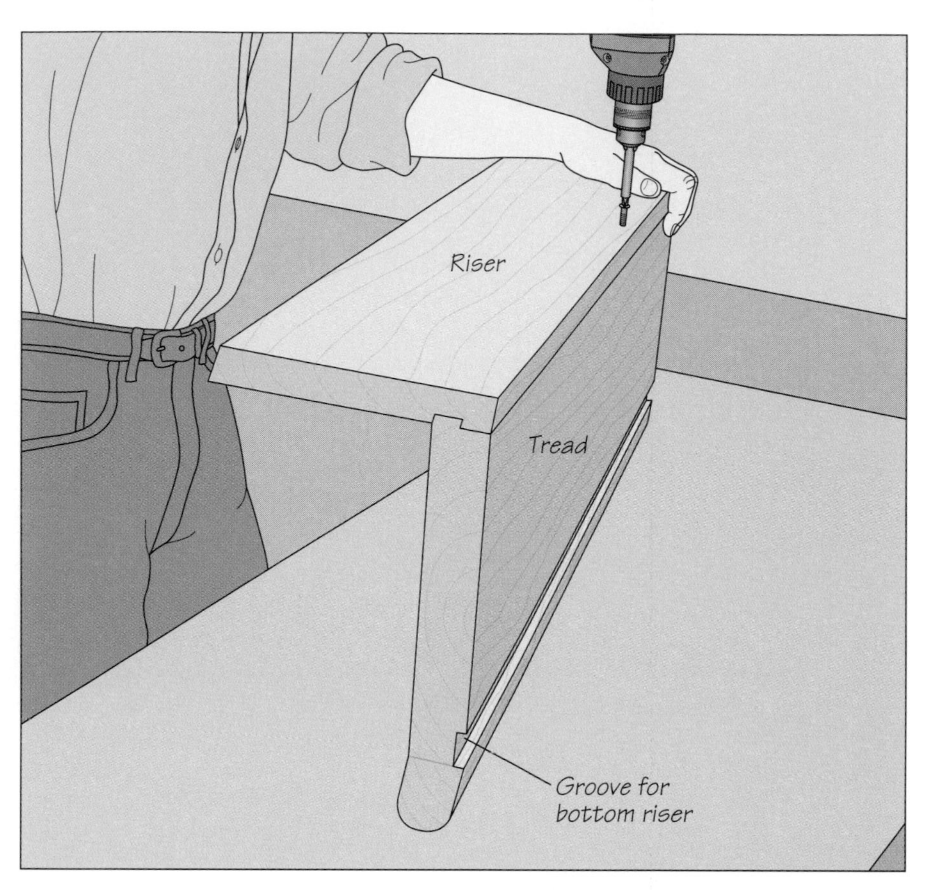

2 Assembling the remaining treads and risers

Once the bottom riser is in place, the remaining treads and risers are assembled and installed in pairs. If you are using dovetails to join the balusters to the treads, first cut the sockets in the treads *(page 127)*. To join a tread and a riser, spread glue in the rabbet at the back of the tread and in the groove in the riser and fit the boards together. Then, holding the tread and riser on a work surface as shown, drive a screw every 3 or 4 inches through the back of the riser *(left)*.

3 Installing the treads and risers

Once all the treads and risers have been assembled, install them one at a time starting at the bottom of the stairs. Apply glue in the groove in the underside of the tread and fit it over the last riser installed. Use finishing nails to fasten the riser to the open and center stringers. Then screw the tread to both the open and center stringer. Counterbore the screw holes so that you can cover the screw heads with wood plugs. Tap a wedge into the closed stringer under each tread and behind every riser *(right)*. You may have to cut some of the wedges short to fit adjoining ones in place. Glue and nail two or three glue blocks to the underside of each tread-riser joint *(page 123)*; locate the blocks near the stringers. Finally, glue and nail the cove molding in place and set all your nails.

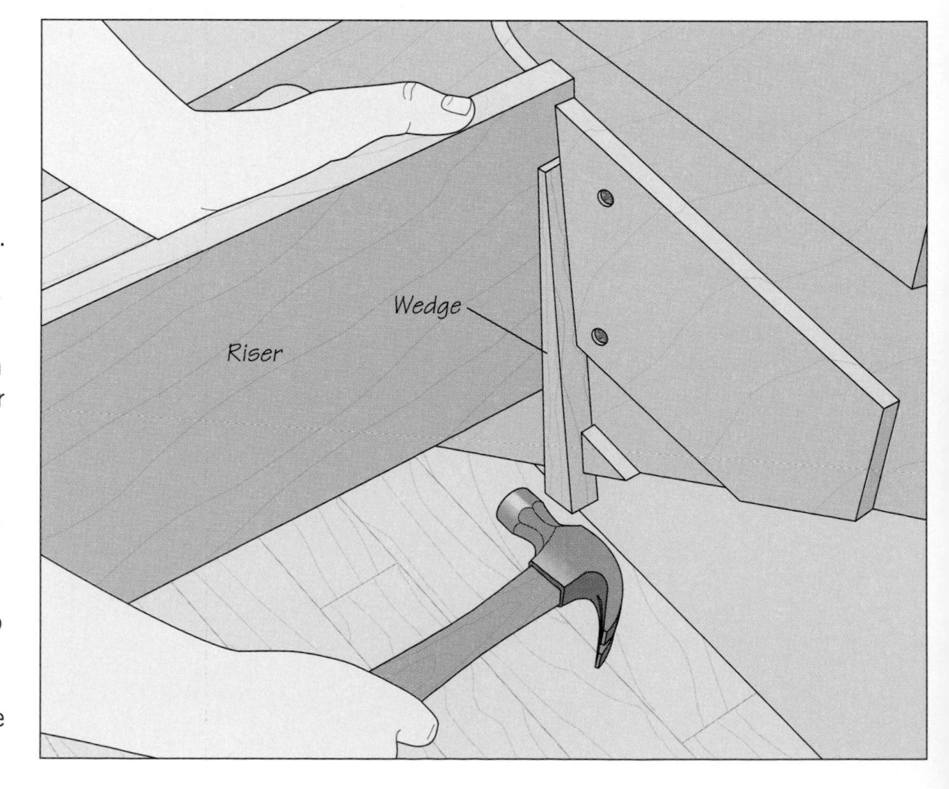

PREPARING TREADS FOR BALUSTERS

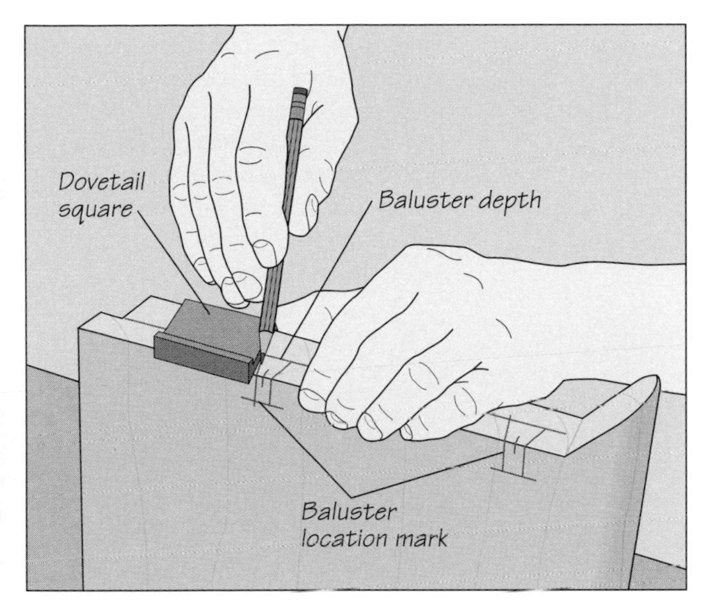

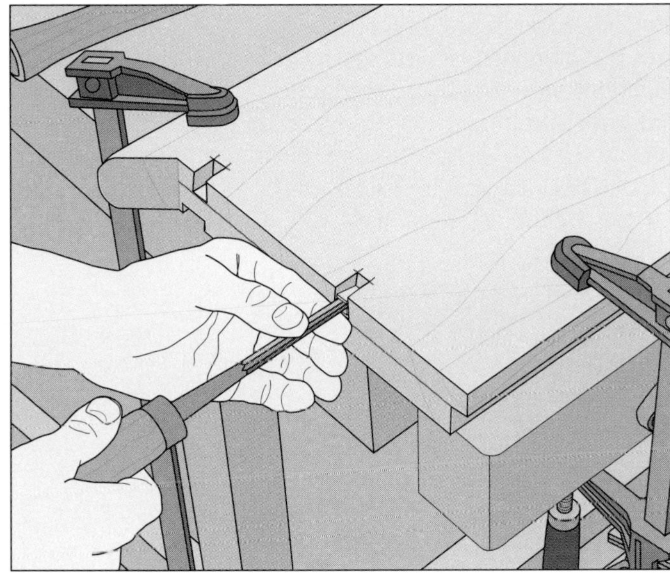

Cutting dovetail sockets

Mark the position of the balusters on each tread *(page 138)* before assembling the treads and risers. Then clamp the tread end up in a vise. Use a dovetail square and a pencil to outline the sockets on the end of the tread at each baluster location mark *(above, left)*. The marks should be centered within the outlines. Extend the socket outlines with straight lines across the top face of the tread about ¾ inch in from the end of the tread. Mark a line on the edge of the tread for the depth of cut. Cut the sockets with a router fitted with a dovetail bit, or use a chisel and a mallet. In either case, clamp the tread face up to a work surface. To cut the sockets by hand, score the outlines with the chisel and a wooden mallet, holding the chisel vertically with the bevel facing the waste. Then pare away the waste in thin layers *(above, right)*, pushing the chisel into the end grain with the bevel facing up.

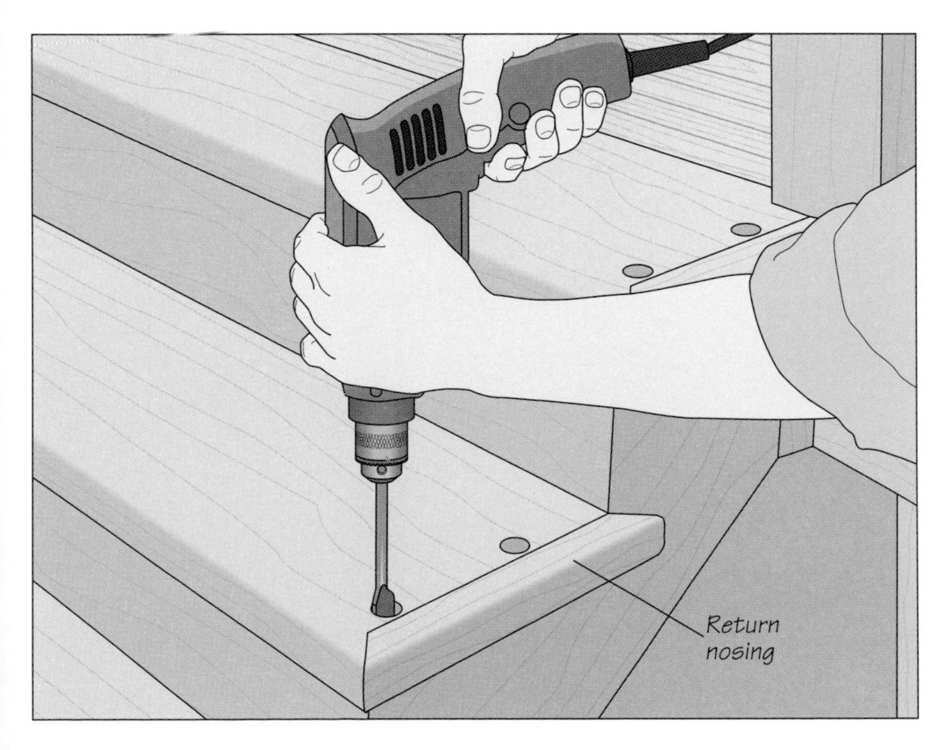

Drilling mortises

If your balusters will be mortised into the treads, you can drill the holes after the treads and risers are assembled. In this case, the return nosings will have already been fastened to the treads *(page 139)*. Turn a tenon on the bottom ends of the balusters *(page 137)*, and mark their positions on the treads *(page 138)*. Fit an electric drill with a spade bit. Holding the drill perfectly straight, bore a hole at each baluster location mark *(left)*, penetrating the tread slightly deeper than the length of the baluster tenon. Locate the holes so that the balusters will be aligned with the center of the newel posts.

NEWEL POSTS

Structurally, newel posts anchor the handrail and balusters of a stairway. But they can also serve as important decorative elements. Newels can be very elaborate structures with boxed enclosures surrounding metal support rods and carved wooden caps or simple pieces of solid or glued-up 4-by-4 stock. And as shown below, wood newels can be turned on a lathe, chamfered or tapered, with many variations possible for each. However complex or straightforward their design, the strength of a newel post derives from how solidly it is attached to the stairs.

The newel post at the top of a staircase is called the landing newel; the one at the bottom is known as the starting newel. As shown on page 129, the joinery attaching each one to the stairs is different from that used for balusters. As a general rule, starting newels should be about 4 feet long. Landing newels range from 5 to 6 feet in length. The precise length of a newel depends on how it is fastened at the bottom. A newel that is bolted to the floor framing will need to be longer.

Instead of a conventional single newel post, the staircase shown above uses a series of several balusters arranged in a tight circle to serve the same visual purpose and to anchor the handrail at the base of the stairs.

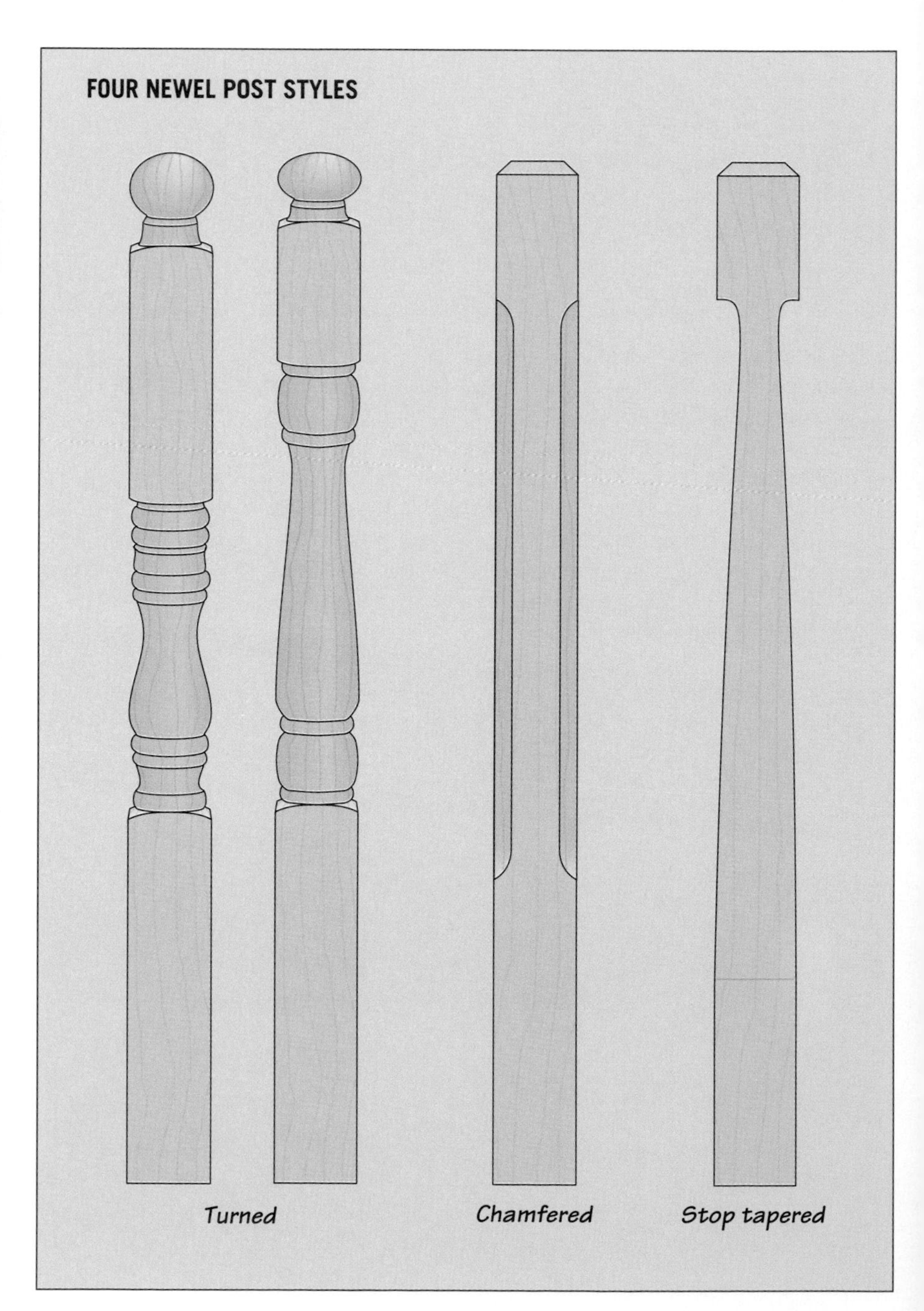

ANATOMY OF STARTING AND LANDING NEWELS

Landing newel
Notched to fit against first riser and tread of stair above landing.

Starting newel
Extends to floor at bottom of stairs; notched to fit around open-stringer end of bottom riser and tread. Could be made longer to extend through subfloor and be bolted to floor joist for added support

TURNING A NEWEL POST

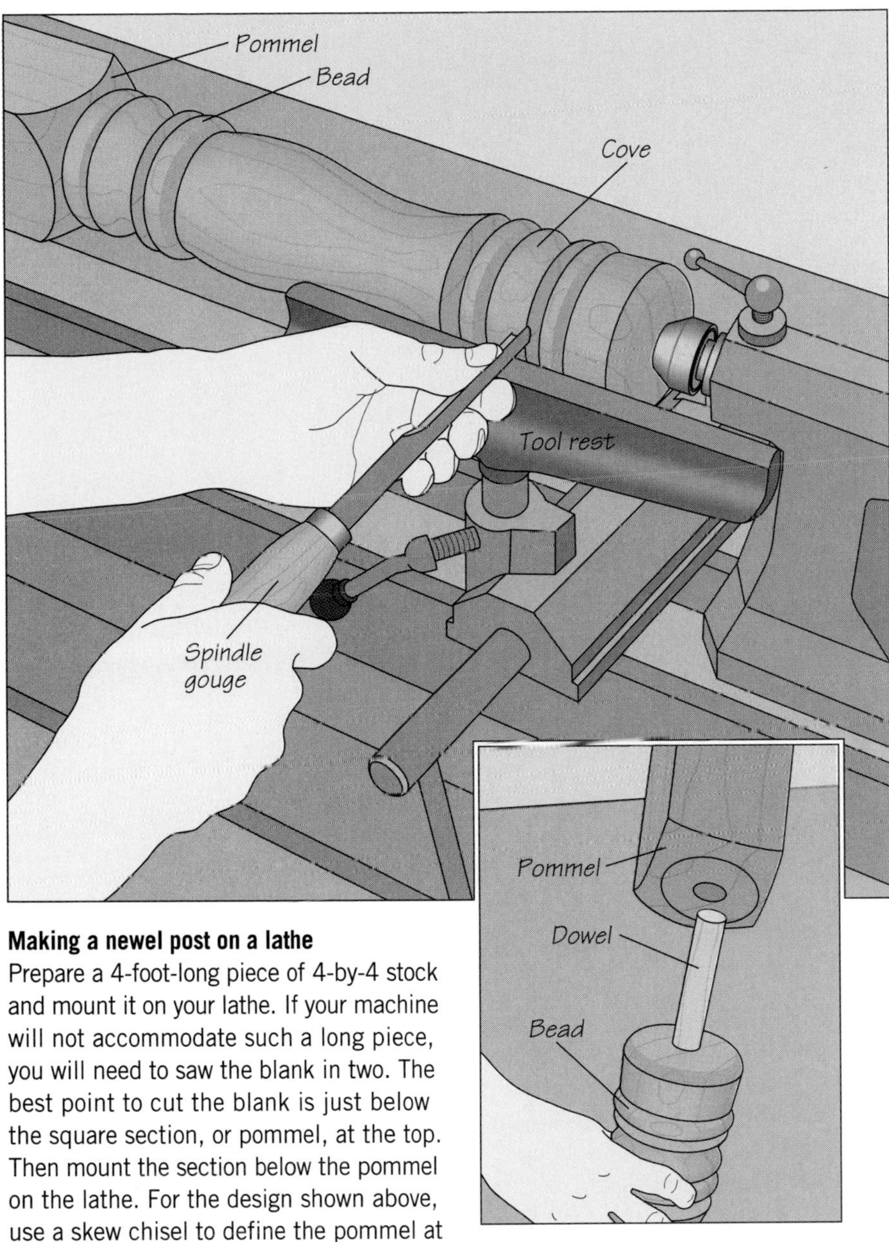

Making a newel post on a lathe
Prepare a 4-foot-long piece of 4-by-4 stock and mount it on your lathe. If your machine will not accommodate such a long piece, you will need to saw the blank in two. The best point to cut the blank is just below the square section, or pommel, at the top. Then mount the section below the pommel on the lathe. For the design shown above, use a skew chisel to define the pommel at the bottom of the post, making a V-cut directly above it. Use a roughing gouge to cut away the bulk of the waste from the rest of the blank and turn it into a cylinder. Switch to a skew chisel or spindle gouge to turn the beads and a spindle gouge for the coves *(above)*. For all these cuts, keep the bevel of the cutting tool rubbing on the stock at all times while bracing the blade on the tool rest. Once the bottom part of the post is turned, mount the pommel at the top on the lathe and make a rounded bevel cut at its bottom end. To rejoin the two sections of the post, install a Jacobs chuck in the lathe tailstock and bore a dowel hole about 3 inches into the adjoining ends of each piece. Cut a 6-inch length of dowel, apply glue into the holes, and insert the dowel. Press the two sections together *(inset)* and clamp. Turn the other newel post the same way.

CHAMFERING A NEWEL POST

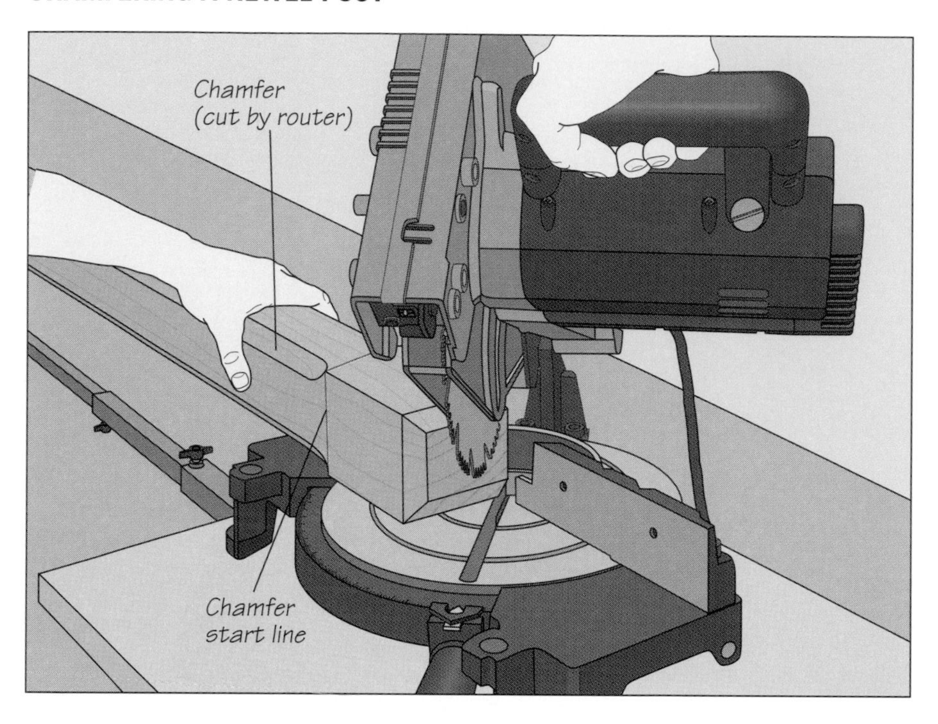

Using a router and a power miter saw
Define the square sections at the top and bottom of the newel post by marking lines across all four sides of the blank. To cut the decorative chamfers along the corners of the post, clamp the blank to a work surface and use a router fitted with a piloted chamfering bit. Stop the cuts at your marked lines. Once all four corners are shaped, chamfer the top of the post on a power miter saw. Set the blank on the saw table and adjust the blade to a 45° angle. Cut off a small wedge of wood, rotate the blank by 90° on the table and repeat the bevel cut. Repeat twice more to finish chamfering the top of the post *(left)*.

INSTALLING THE NEWEL POSTS

1 Preparing the starting newel
Position the starting newel on the corner of the bottom tread so that the middle of the post is in line with the dovetail sockets or mortises for the balusters. Use a pencil to mark cutting lines on the bottom of the post so it will butt against the riser and open stringer. Extend the lines up the inside faces of the post. To determine where to stop these cuts so that the post rests on the floor, measure from the top of the bottom tread to the floor and transfer your measurement to the post, marking cutting lines on its inside faces. Cut into the post along these lines with a handsaw, stopping the cuts at the first set of cutting lines. Make the remaining cuts on your table saw. To set up the saw, set the post on the saw table and raise the blade to the horizontal cutting line. Then align the front end of the blade with the handsaw cuts you made, butt a board against the end of the post and clamp it to the rip fence as a stop block. Finally, align the vertical cutting line on the end of the post with the blade and butt the fence against the stock. Feed the post with both hands, running it against the fence until it contacts the stop block. Rotate the post 90°, reposition the fence, if necessary, and repeat the cut *(right)*. Use a chisel and a mallet to detach the waste piece from the post. Test-fit the post on the tread; to ensure it overhangs both tread and riser by the same amount, you will have to trim the tread nosing. This cut is shown in step 3 *(page 131)*. **(Caution: Blade guard removed for clarity.)**

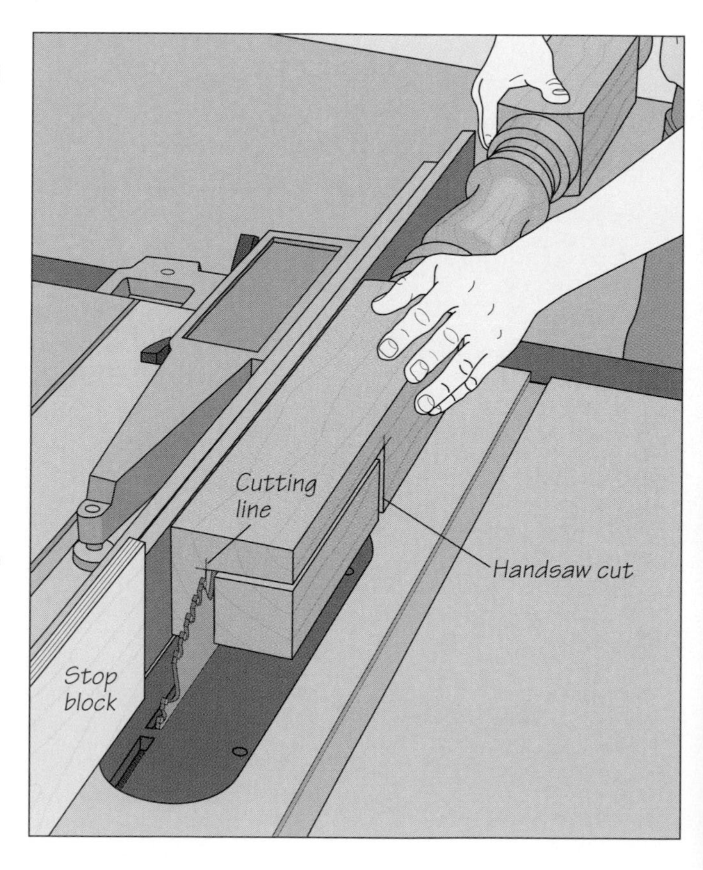

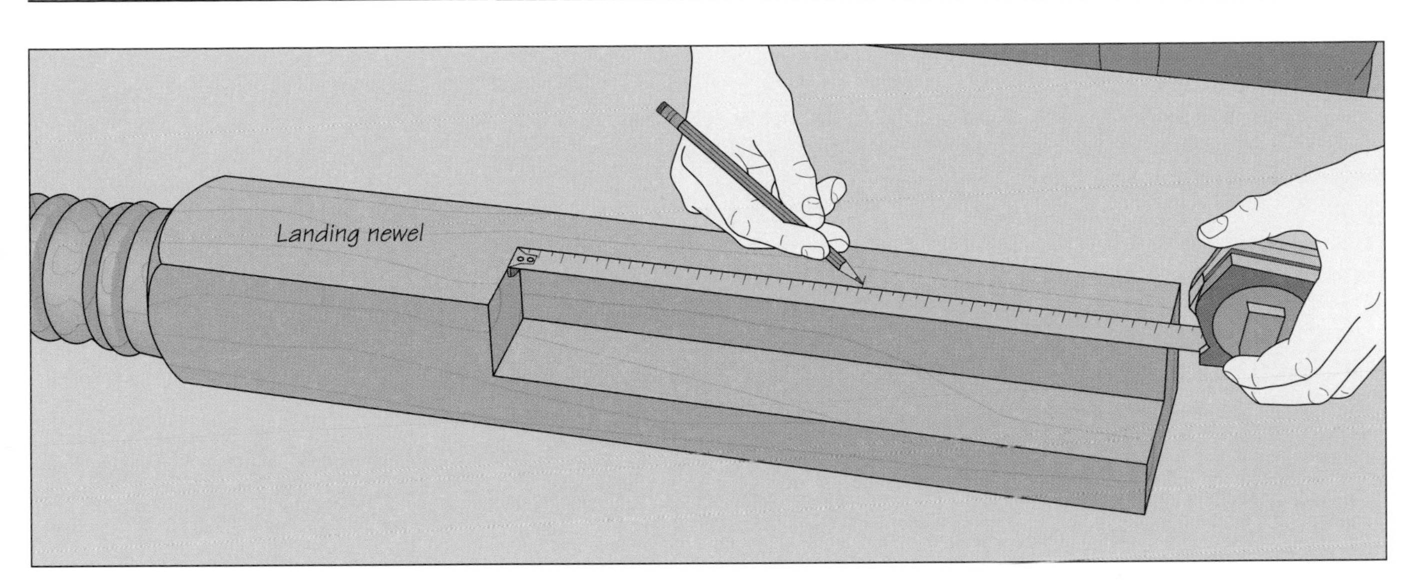

2 Preparing the landing newel

The landing newel is prepared the same way as the starting newel, except that it is 6 to 8 inches longer, since it will extend to the open stringer for support. As a result, you will need to make one additional cut. Measure the cut-out you made in the starting newel and lengthen the measurement by the difference in length between the two posts. Mark cutting lines on the landing newel and notch it as you did the starting newel *(page 130)*. Be sure the tops of the newels will be the same height when the posts are installed. Now, measure the gap between the landing and the tread directly above it, and transfer the measurement to the face of the post that will extend to the landing *(above)*. Cut the newel with a handsaw. A decorative bevel on the bottom of the post will give it a lighter, more finished appearance.

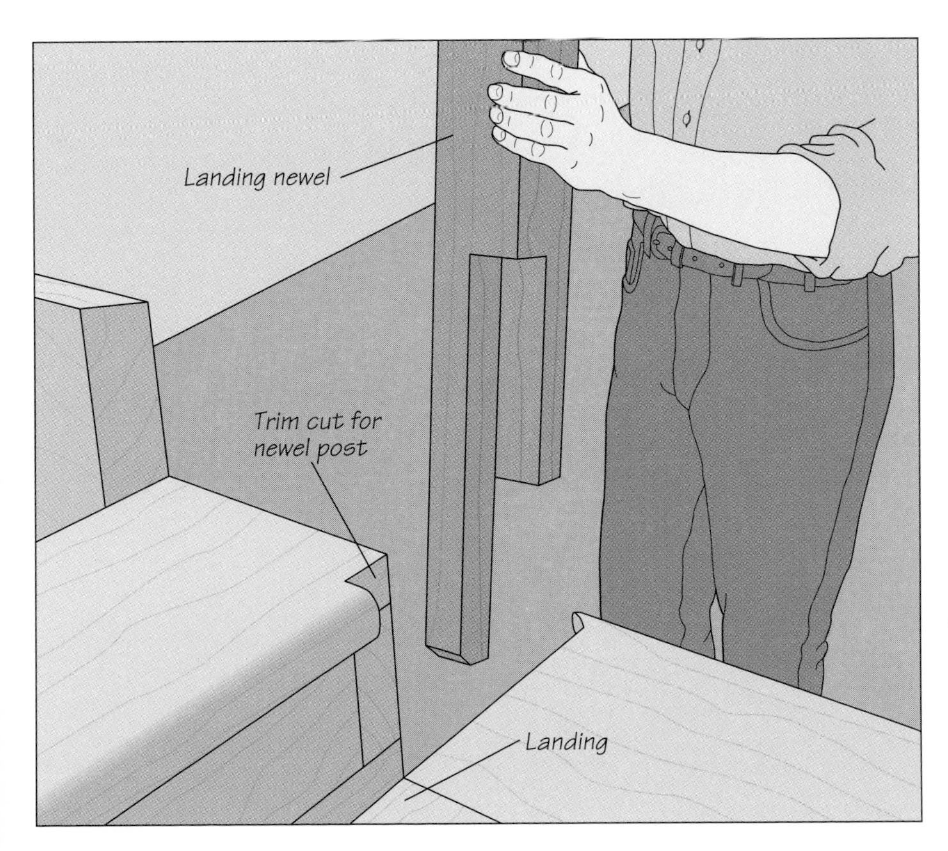

3 Installing the newel posts

The newels will be fastened to the stringers with lag bolts, driven into adjoining sides of the posts. With a helper holding each newel in position on the stairs, work from the inside of the staircase to drive two lag bolts through the stringer and into the post.

HANDRAILS

A handrail can be the most complex and decorative element of a staircase. But it also performs the more pedestrian—but vital—task of guiding the people who climb and descend the stairs. Whether a handrail is as elaborate as a curved assembly made from laminated strips of wood, or as simple as the straight example featured in this section, most building codes govern several aspects of its construction. For example, a handrail is usually required on any staircase with three or more treads. It is typically screwed or bolted to the newel posts and attached to the tops of the balusters. A rail should also typically not encroach more than 3½ inches into the minimum width of the staircase. Most codes require stairs wider than 44 inches to have a handrail on both sides. A handrail along a wall is called a floating handrail—so-called because it is suspended above the treads, attached to the wall above them *(page 135)*. Commercial rails and hardware for floating handrails generally satisfy building codes.

This section will show you how to make and install a handrail along the open-stringer side of a staircase as well as how to mount a floating handrail. The designs shown below can be made on a shaper or router table using 2-by-3 stock. The groove along the bottom of the rail is cut on the table saw and houses the tops of the balusters.

The handrail shown at right culminates in a spiral-shaped form, known as a volute, at the foot of the stairs.

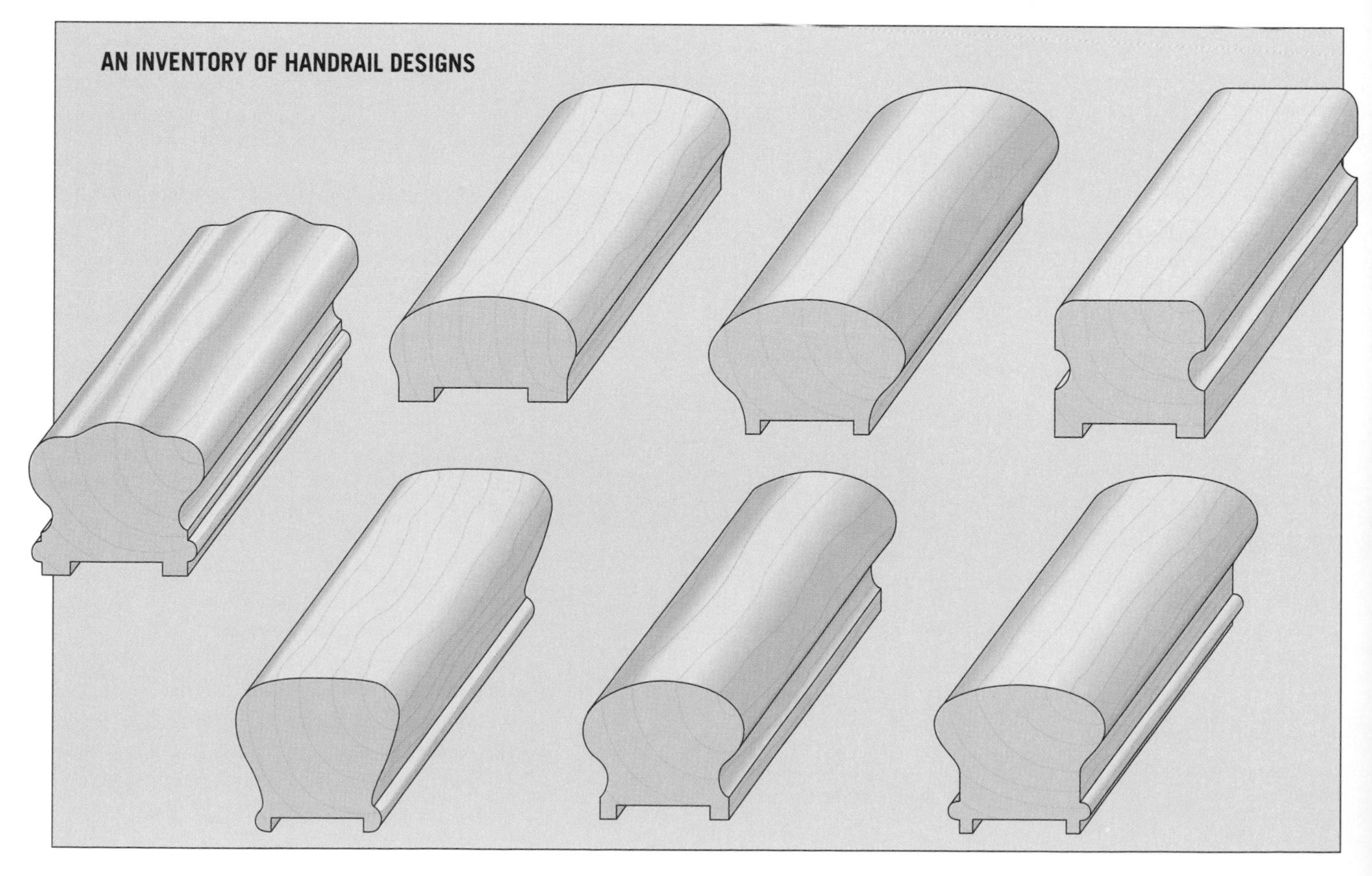

SHAPING A HANDRAIL

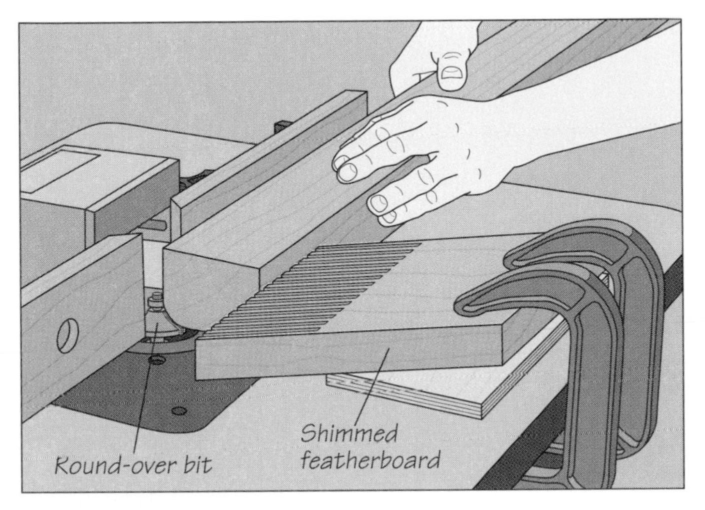

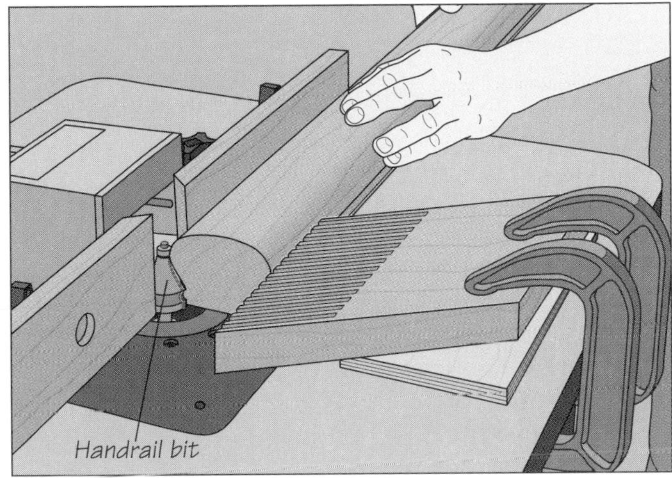

1 Routing the profile
Install a bit with the desired top profile in your router and mount the tool in a table. In this example, a round-over bit is shown. Align the fence with the bit pilot bearing. To support the handrail as you shape it, clamp a featherboard to the table in line with the bit. Place a shim under the featherboard so that the pressure is applied near the middle of the stock. Feed the rail upside down, then turn it around and shape the other edge of the top *(above, left)*. To form the side profile, replace the bit. In this case, a specialized handrail bit is used. Feed the rail across the table in two passes again, this time right side up *(above, right)*.

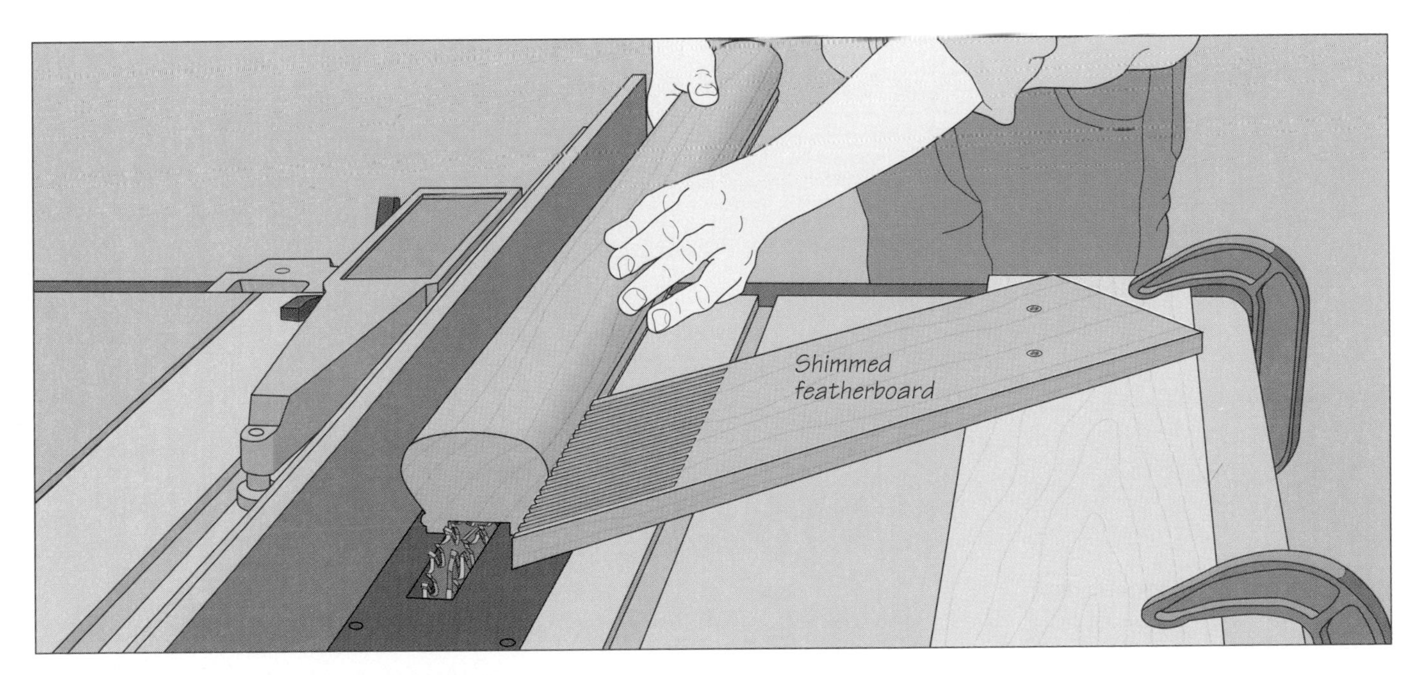

2 Cutting the baluster groove in the handrail
To accommodate the top ends of the balusters, saw a groove down the middle of the handrail's underside. Install a dado blade, adjusting it as wide as possible; you will likely need more than one pass to cut the full width of the groove. Adjust the cutting height to about ¼ inch, then mark the groove in the center of the leading end of the stock. The width of the groove should be equal to the thickness of the balusters. Clamp a shimmed featherboard to the saw table in line with the dado head. Align one of the groove marks with the inside blade, butt the rip fence against the handrail and feed the stock into the cut. Turn the handrail around and repeat the cut *(above)*.

INSTALLING THE HANDRAIL

1 Cutting the handrail to length
Once you have shaped the handrail, position it on the treads, butting an edge against the newel posts at the top and bottom of the stairs. Holding the rail in place, mark lines across its edge at both points where it meets the newels. Depending on the method you select to attach the handrail to the starting newel post *(below)*, you can make the rail slightly longer, since you may need to mortise the bottom end of the rail into the post. Then adjust a sliding bevel so that its handle is parallel to the rail and its blade is flush on the landing newel *(right)*. Use the bevel's setting to set up your saw to the proper angle for cutting the handrail, then cut the rail to length.

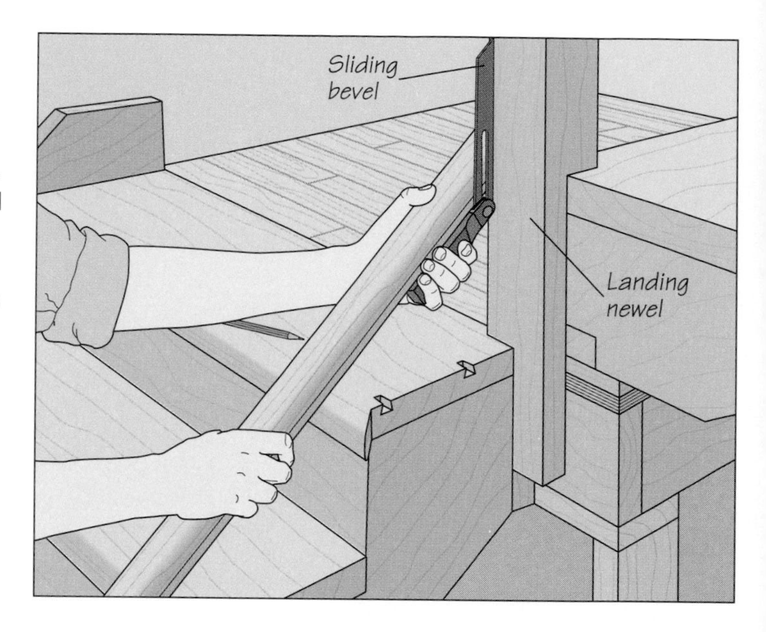

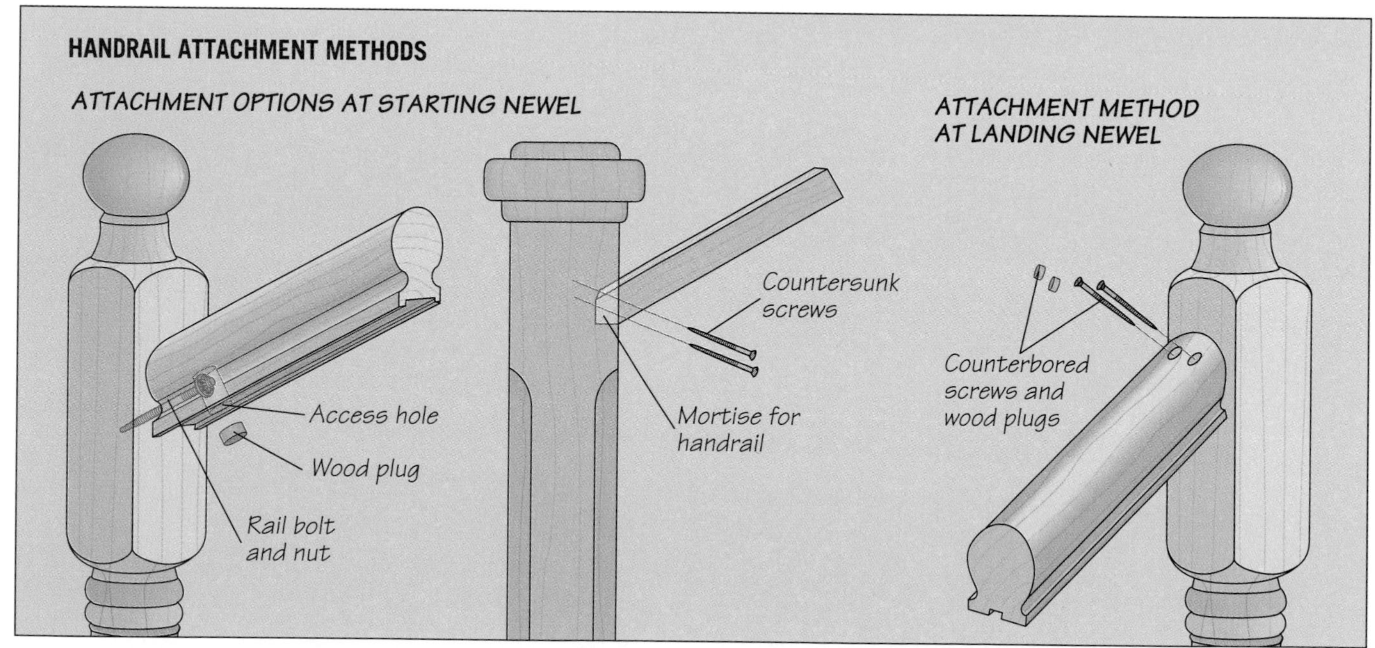

2 Choosing an attachment method
There are several ways of fastening a handrail to the newel posts; three methods are shown above. Most finish carpenters rely on one of two options for the starting newel. The traditional method involves using a rail bolt. A clearance hole for the bolt is drilled into the bottom end of the rail and a pilot hole is bored in the newel; an access hole is also drilled into the bottom face of the rail. The bolt is then driven partway into the newel and the handrail is slipped over the protruding end of the bolt. A special nut is then inserted through the access hole and onto the bolt, and tightened using a small screwdriver. A wood plug is glued in the access hole to conceal the bolt. The second method involves cutting a mortise in the newel for the end of the rail. The rail is fitted in the mortise and screwed to the newel. The best option at the landing newel involves simply butting the end of the rail against the post and screwing it in place. In this case, since the screws are driven into the top face of the rail, the holes are counterbored and the heads are covered with plugs. Screws are shown in the step that follows.

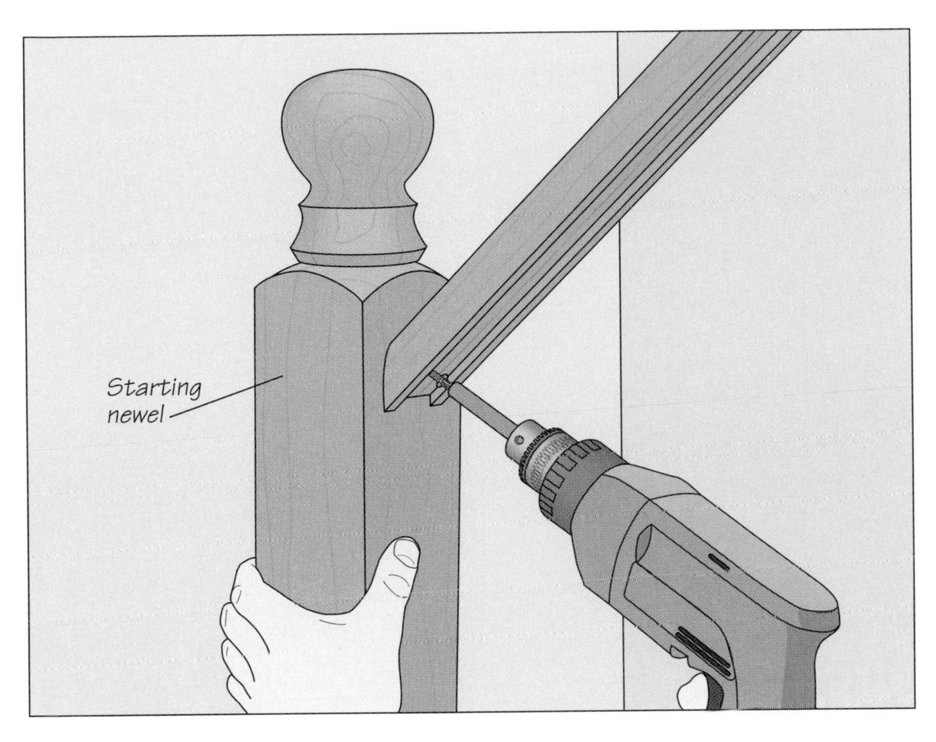

3 Attaching the handrail to the newels
To fasten a handrail to newel posts with screws, hold the rail in position and drill a clearance hole though the rail and a pilot hole into the posts. At the top of the stairs, drill counterbored holes and drive the screws through the top face of the rail. Conceal the screw heads with wood plugs. At the bottom, work from the underside of the rail and countersink the screws *(left)*.

INSTALLING A FLOATING HANDRAIL

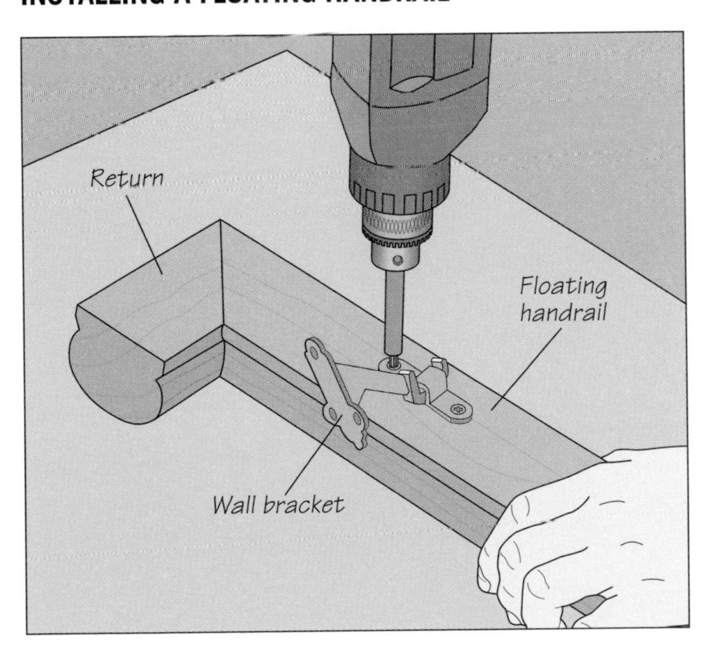

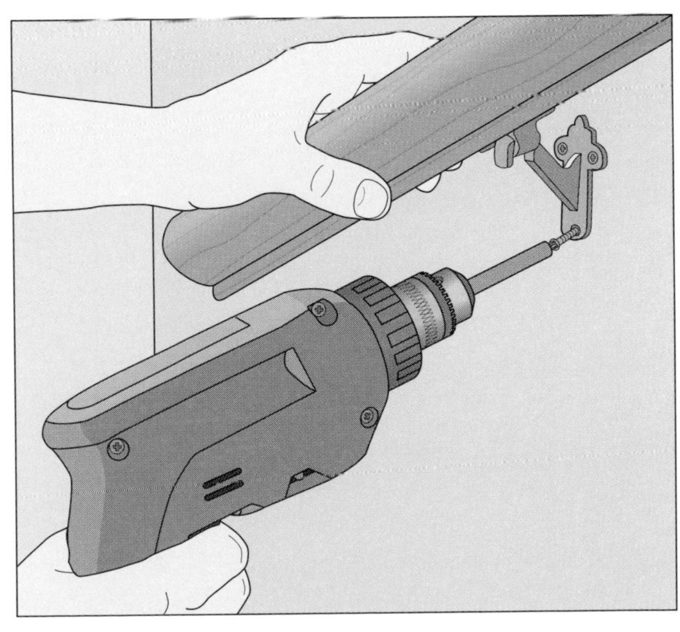

Attaching the handrail
Make the floating handrail *(page 133)*, omitting the groove in the underside, and cut the handrail to the same length as the handrail on the open-stringer side of the staircase. The model shown above features a return that serves as a tactile reminder to the visually impaired that they are arriving at the top or bottom of the stairs. Locate the studs along the wall side of the stair. Then position the floating handrail against the wall parallel to the other handrail and mark the stud locations on it. Screw commercial wall brackets to the underside of the rail *(above, left)* at the stud location marks at intervals specified by your local building code. Reposition the handrail on the wall, mark the screw holes, bore pilot holes into the wall, and fasten the rail in place *(above, right)*.

BALUSTERS

Like newel posts and handrails, balusters can be made in a wide variety of designs. Four popular styles are illustrated at right. Although balusters can be bought ready-made, they can be fashioned easily in the shop. On the following page are instructions for turning balusters on a lathe and making tapered balusters on a jointer.

As shown in the illustration on page 138, not all the balusters are cut to the same length. The baluster at the back of each tread is longer than the one at the front, owing to the slope of the handrail. The square section at the top of the balusters—if there is one—is typically the same length on all of them, but not that at the bottom. Most balusters are cut to the length of the gap between the handrail and treads, adding about ¾ inch for the tenon or dovetail you will use to attach the balusters to the treads. Balusters with rounded top sections should be longer to accommodate the tenon that fits into the handrail mortise.

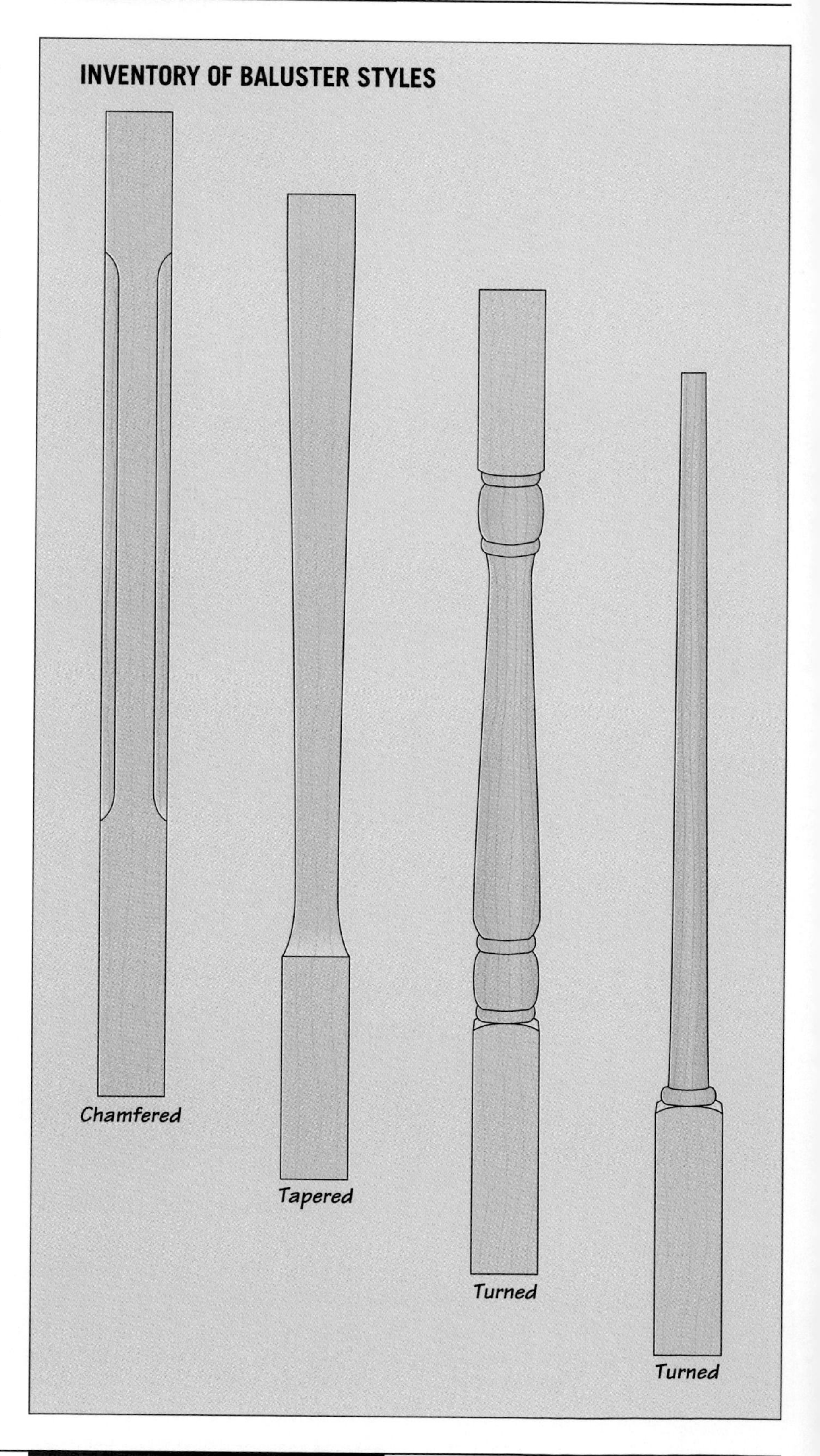

MAKING BALUSTERS

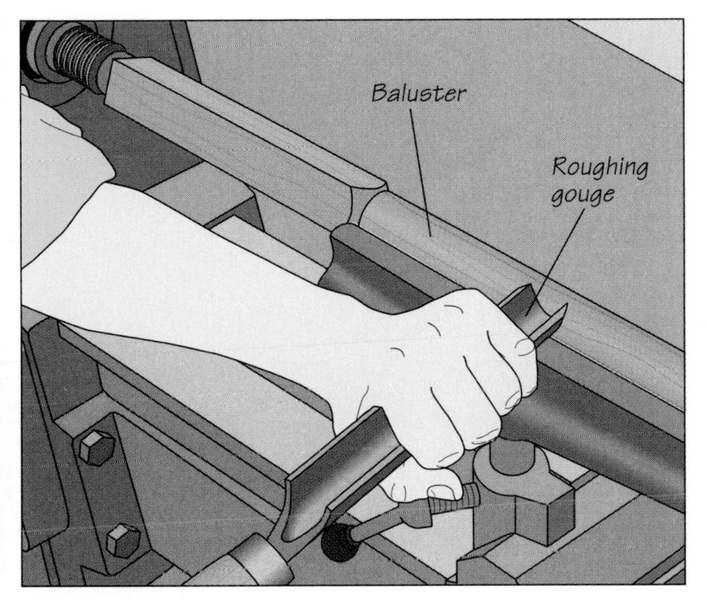

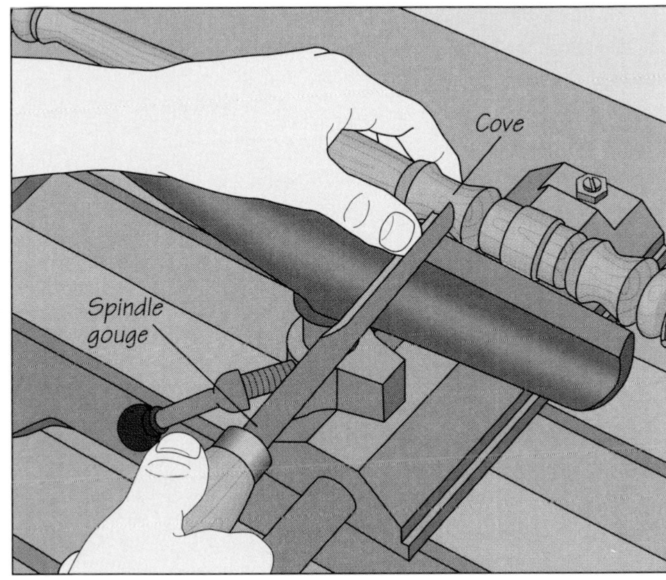

Turning balusters on the lathe

Mount a baluster blank on your lathe and mark off the square sections at the top and bottom. Define these sections as you did for the newel posts *(page 129)*, then use a roughing gouge to turn the blank into a cylinder between the square areas *(above, left)*. Turn any other design elements, such as beads or coves, using a spindle gouge and a skew chisel. When turning relatively thin sections, support the blank with your left hand to prevent chatter *(above, right)*. If you are using tenons to join the balusters to the treads, turn a ¾-inch-long tenon at the bottom of each baluster using a parting tool. The diameter of the tenons should match the mortises cut in the treads *(page 127)*.

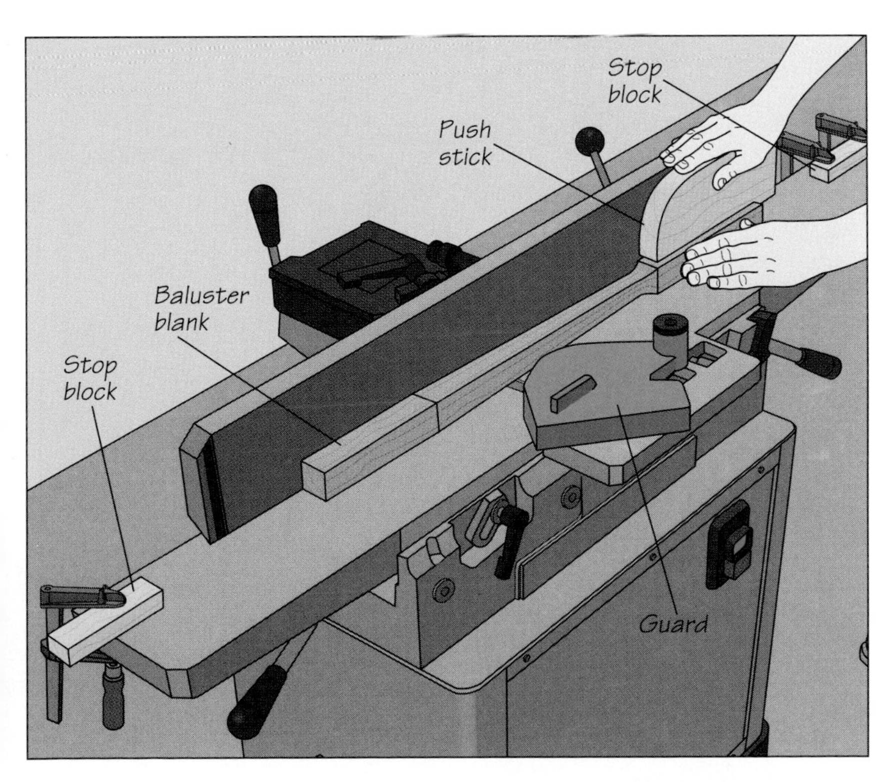

Making a tapered baluster

Move the guard out of the way for this operation and adjust the jointer for a ⅛ inch cut. Mark off the square sections at the top and bottom of each baluster blank and set the blank on your jointer, aligning the mark for the top square section with the front edge of the outfeed table. Butt a wood block against the end of the stock and clamp it to the outfeed table as a stop. Repeat with the other mark to clamp a stop block to the infeed table. To make the first pass, butt the end of the blank against the infeed stop block and lower the blank onto the knives, keeping it flush against the fence. Feed the blank with a push stick, using your left hand to press the workpiece against the fence until it contacts the outfeed stop block, then lift it off. Make one pass on each side of the blank, then increase the cutting depth by ⅛ inch and repeat the process on all four sides. Continue, increasing the cutting depth until you obtain the desired taper.

INSTALLING THE BALUSTERS

1 Planning the operation
Building codes do not provide many rigid guidelines governing baluster installation, but most codes require them to be no more than 6 inches apart. As shown at right, stairs typically feature two balusters per tread; for visual balance, space them an equal distance apart. The front one on each tread is typically positioned just behind the nosing. The back baluster is then positioned halfway between the front baluster and the front baluster on the tread above.

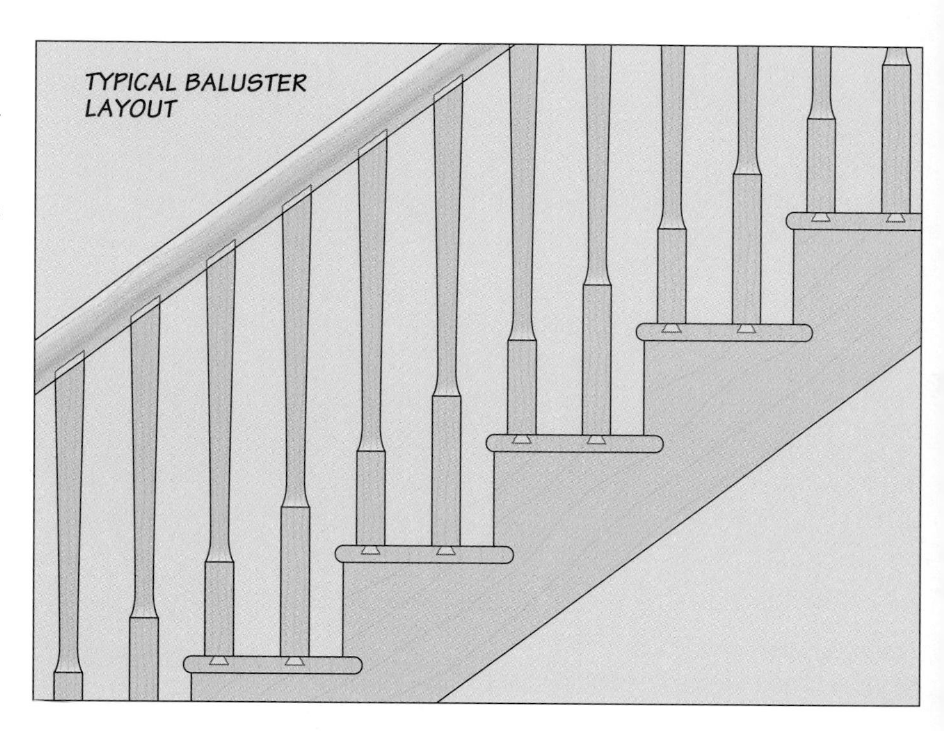

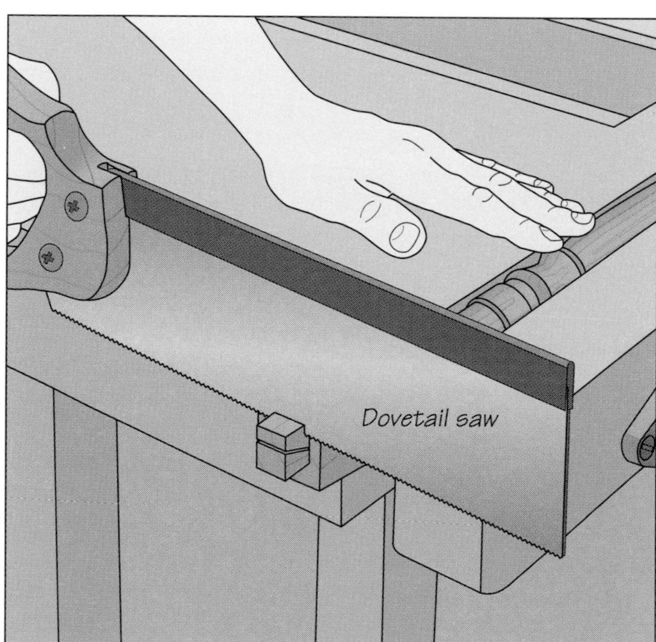

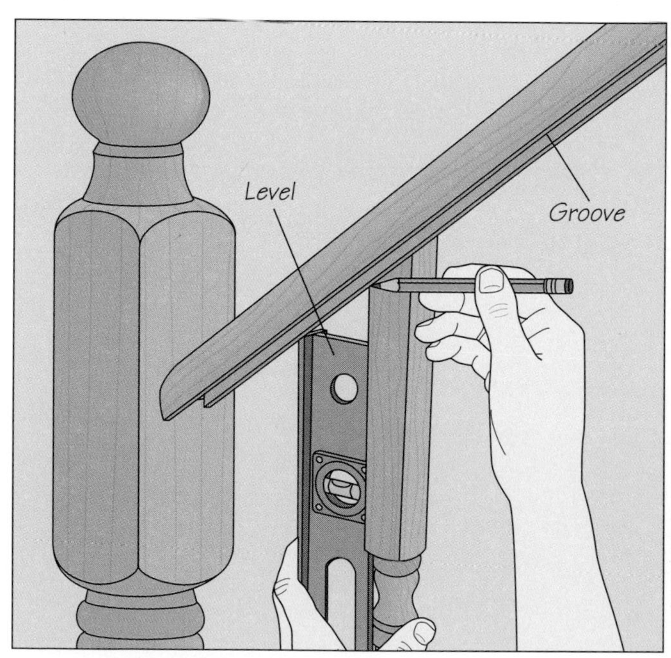

2 Cutting dovetails on the balusters
For balusters that will be dovetailed into the treads, position a baluster up against the end of the tread and outline the socket you cut in it on the bottom of the baluster. Then secure the baluster in a vise and cut the dovetail using a dovetail saw *(above)*. Use the finished baluster as a template to outline the remaining ones.

3 Cutting balusters to length
Fit the bottom of the baluster into its tread and, holding it perfectly upright against the handrail, mark a line on it along the underside of the rail *(above)*. Mark a second line above the first, offset from it by the depth of the groove in the rail. Saw the balusters to length along this second marked line.

4 Fastening the balusters in place
Once all the balusters are cut to length, install them one at a time. Position the baluster between the tread and the handrail, and drill two pilot holes for finish nails through the top end of the baluster into the underside of the rail. Glue the bottom of the baluster to the tread. If you are using dovetails, also drive a nail through the dovetail and into the tread; for tenoned balusters, the adhesive is adequate. Then add glue to the top end of the baluster, butt it against the hand-rail, and nail it in place *(right)*. To hide the gaps in the handrail groove between balusters, cut wood strips about ¼ inch thick, called fillets. Glue and nail them in place.

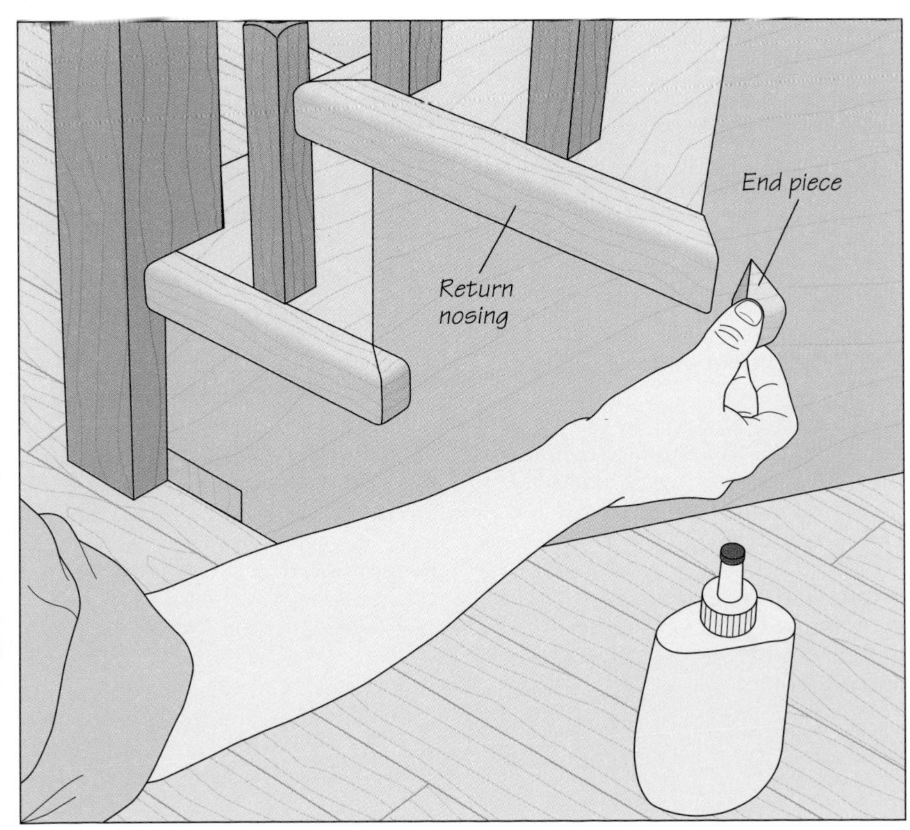

5 Finishing up the job
Once all the balusters are in place, complete the staircase by installing return nosings to cover the end grain of the treads on the open-stringer side. Make the nosings by ripping them from rounded-over tread stock; the width of the pieces should be the same as the overhang of the tread nosings from the risers *(page 125)*. Saw the return nosings to extend to the open stringer, mitering the ends to match the miter you cut on the treads. Glue and nail the nosings in place and set the nails. To conceal the end grain of the return nosings, cut end pieces with one 45° angle and glue them to the nosings *(left)*. Hold the end pieces in place with masking tape until the adhesive cures.

GLOSSARY

A-B-C

Apron: In stool-and-apron window casing, a horizontal piece of window trim installed beneath the stool.

Baluster: A vertical post mounted between the handrail and treads opposite the wall side of a staircase.

Baseboard: Decorative trim installed around the perimeter of a room at the base of the walls.

Bead: A convex profile, usually semicircular.

Bed molding: A type of crown molding featuring reversed curves; often used as part of a formal cornice.

Bevel cut: Sawing at an angle from face to face through the thickness or along the length of a workpiece.

Built-up baseboard: A type of baseboard built up from several elements, such as base-and-shoe baseboard.

Cap rail: A molded rail installed at the top of wainscoting.

Casing: Decorative trim used to frame a window or door.

Chair rail: Decorative trim installed on walls, usually about 3 feet above the floor. Traditionally served to prevent chair backs from damaging walls.

Closed stringer: A 2-by-10 or 2-by-12 that is mortised to accept and support the ends of the treads and risers of a stairway; usually adjacent to a wall. See *open stringer.*

Compound cut: A cut through a board at angles other than 90° relative to the face and edge of stock.

Cope-and-stick joinery: A method of joining stiles and rails in frame-and-panel doors and windows. Cut with a router or shaper, the joint features mating tongues and grooves and a decorative molding along the inside edges of the boards.

Coped joint: A method of joining two pieces of molding at an inside corner. The end of one board is cut so that it fits precisely against the contoured face of the mating board.

Countersinking: Drilling a hole that enables a screw head to lie flush with or slightly below the surface.

Cove molding: Trim featuring a concave profile.

Crosscut: Sawing across the grain of a workpiece.

Crown molding: Decorative trim installed around the perimeter of a room at the top of the walls; also known as cornice molding.

D-E-F-G-H-I

Dado: A rectangular channel cut into a workpiece.

Double-hung window: A type of window consisting of two sashes that slide vertically within a jamb.

End grain: The arrangement and direction of the wood fibers running across the ends of a board.

Formal cornice: An elaborate type of crown molding built up from a box-like soffit-and-fascia assembly decorated with strips of crown and bed molding.

Frame-and-panel door: A door consisting of panels contained within a framework of stiles, rails, and mullions.

Furring strip: A narrow strip of wood nailed to wall studs to support molding or wainscoting.

Glass-stop molding: Decorative strips of wood used to hold a pane of glass in a window sash.

Glazing bar half-lap joint: A method of joining the rails and mullions of a window sash with mitered half-laps.

Glue block: In stairbuilding, a triangular piece of wood glued and nailed under the joint of a tread and riser to reinforce the joint between them.

Gooseneck: A nearly vertical piece of railing connecting a handrail to the upper newel post of a staircase.

Half-blind dovetail: A joinery method involving interlocking pins and sockets; commonly used to join balusters to treads.

Hinge jamb: The side of a door jamb adjoining the hinge side of the door.

J-K-L-M-N-O-P-Q

Jamb extension: A wooden frame installed on a window jamb to bring it flush with the interior wall.

Joist: A horizontal support for a floor.

Kicker: A 2-by-4 attached to the subfloor to anchor the stringers at the bottom of a staircase.

King stud: A vertical framing member adjoining the rough opening for a door.

Ledger board: A 2-by-4 attached to the headboard at the top of a staircase to support the stringers.

Level: Horizontal; parallel to the floor or ceiling. See *plumb.*

Miter cut: A cut that angles across the face of a workpiece; see *bevel cut.*

Molding head: A solid metal wheel installed on the arbor of a table saw or radial arm saw for forming moldings; holds three identical knives.

Mortise-and-tenon joint: A joinery technique in which a projecting tenon on one board fits into mortise in another.

Mortise: A hole cut into a piece of wood to receive a tenon.

Mullion: A vertical member between two rails of a frame.

Nailer: A 2-by-4 installed horizontally between wall studs to support wall paneling.

Newel post: A wooden post fastened to the handrail and treads at the top and bottom of a staircase.

Open stringer: A 2-by-10 or 2-by-12 that is notched to support the ends of the treads and risers of a stairway; usually away from a wall. See *closed stringer.*

Picture rail: Decorative trim installed on walls, usually about 6 feet above the floor, for hanging picture frames.

Pilot hole: A hole bored into a workpiece to prevent splitting when a screw is driven; usually made slightly smaller than the threaded section of the screw.

Pilot bearing: A free-spinning metal collar on a piloted router bit that follows the edge of the workpiece or a template to keep the cutting depth uniform.

Plinth: A decorative wood block installed between the side casing of a door and the floor.

Plumb: Vertical; perpendicular to the floor and ceiling. See *level.*

Preacher: A U-shaped jig used to mark the length of a piece of molding that butts against door trim or a plinth.

R-S

Rabbet: A step-like cut in the edge or end of a board; usually forms part of a joint.

Rail: The horizontal member of a frame-and-panel assembly; see *stile.*

Return nosing: A piece of tread stock nailed and glued to the open-stringer end of a tread to conceal the end grain of the tread and the bottom ends of the balusters.

Reveal: The gap between the inside face of a window or door jamb and the inside edge of the trim installed on it.

Rip cut: A saw cut that follows the grain of a workpiece.

Rise: The vertical distance between two adjoining steps on a staircase; also called unit-rise. See *run.*

Riser: A board that closes the vertical space between stair treads.

Rosette: A decorative wood block installed at the upper corners of window or door casing.

Rough opening: The wall opening into which a window or door jamb is installed.

Run: The horizontal span of each step of a staircase; also known as unit-run. See *rise.*

Scarf joint: A method of joining two lengths of molding end to end by beveling both pieces.

Sole plate: A horizontal framing member installed on the subfloor to support the wall studs.

Stair button: A commercial jig attached to a carpenter's square to lay out the rise-and-run of a staircase on the stringers; usually used in pairs.

Stile: The vertical member of a frame-and-panel assembly. See *rail.*

Stool: The horizontal component of stool-and-apron casing that juts out and forms the sill of the finished window.

Wall stud: A vertical member forming walls and supporting the framework of a building.

T-U-V-W-X-Y-Z

Top plate: A horizontal framing member installed along the top of the wall studs.

Tread: Forms a step of a staircase.

Tread nosing: The rounded front edge of a stair tread.

Volute: A spiral section of a staircase handrail, usually ending at the newel post.

Wainscoting: Wall paneling that covers the lower part of a wall.

INDEX

Page references in *italics* indicate an illustration of subject matter. Page references in **bold** indicate a Build It Yourself project.

A-B-C-D
Balusters, *136*
Chamfering, *137*
Installation, *138-139*
Tread preparation, *127*
Turning, *137*
Baseboards, *12, 23*
Base-and-shoe, *24*
Built-up, *25*
Installation, *26*
inside corners, *26-27*
outside corners, *28-29*
store-bought corner pieces (Shop Tip), *29*
Joinery, *23, 26-29*
Bevel cuts, *18*
Doors, *102, 106*
Build It Yourself:
Doors
door bucks, **104**
Windows
reveal gauges, **63**
Butted sill casings, *65*
Cap rails, *38*
Carpenter's squares, *front endpaper*
Trueing, *front endpaper*
Casings:
Doors, *85, 86, 112-113*
Windows, *12*
butted sill, *65*
correcting poor-fitting miters, *66-68*
picture-frame, *58, 61-65*
stool-and-apron, *57, 58, 69-72*
temporary brace to hold window aprons (Shop Tip), *72*
Ceilings:
Paneled, *52-55*
Chair rails, *12, 30-32*
Coffered ceilings, *52-55*
Combination planes, *25*
Compound cuts, *18*
Coping saws, *15*
Crosscutting, *17*
Crown moldings, *12, 30, 33-35*
Formal cornices, *36-37*
Doors, 7, *13, 85-86*
Beveling, *102, 106*
Binding, *102, 106*
Casings, *85, 86, 112-113*
Doorstops, *98-99*
Exterior, 85, 86
Frame-and-panel, *90-94*
Hanging, 85, 86, *105-106*
Hardware, *89*
locksets, *107-111*
Hinges, *103, 105*
butt hinges, *84*
mortises, *100-102*
sizes, *103*
Interior, 85, 86
Jambs, *95-97*
hinge mortises, *100-102*
Styles, *86, 87*

E-F-G-H-I
Eakes, Jon, 8-*9*
Finish nailers, *14, 21*
Frame-and-panel construction:
Ceilings, *52-55*
Doors, *90-94*
Wainscoting, *39, 40, 41, 46-51*
Glass-stop moldings, *79-80*
Glazing bar half-lap joints, *81-83*
Handrails, *132-133*
Floating handrails, *135*
Installation, *134-135*
Hardware:
Doors, *89*
locksets, *107-111*
See also Hinges
Hinges:
Doors, *89*
butt hinges, *84, 89*
mortises, *88, 100-102*
sizes, *103*

J-K-L
Jigs:
Doors
door bucks, **104**
hinge-mortising jigs, *88*
jamb jigs, *88, 95*
Moldings
preachers, *23*
Stairs
calculating rise-and-run, *115*
Windows
auxiliary tables for power miter saws, *68*
mortising jigs for routers, *75*
reveal gauges, **63**
Joinery:
Baseboards, *23, 26-29*
Paneling
cope-and-stick joints, *46, 47-48*
tongue-and-groove, *43*
Windows, *76, 78*
glazing bar half-lap joints, *81-83*
Log-builder's scribes, *42*
Lumber. *See* Woods

M-N-O-P-Q
Miter boxes, *15*
Miter cuts, *17*
Baseboards
inside corners, *26*
outside corners, *28*

Crown moldings, *34-35*
Windows
correcting, *66-67*
Moldings, 8, *12*, 21-*22*
Cap rails, *38*
Chair rails, *30-32*
Crown moldings, *30*, *33-35*
formal cornices, *36-37*
Fitting
scribing, *front endpaper*
Glass-stop moldings, *56*, *79-80*
installing the molding with a hammer (Shop Tip), *80*
Milling, *back endpaper*
Picture rails, 30
Rosettes, *58*, *73-74*
shop-made rosette cutters (Shop Tip), *74*
See also Baseboards; Casings
Mortises:
Door hinges, *100-102*
Nails:
Finishing, *back endpaper*
Newel posts, *128-129*
Chamfering, *130*
Handrail attachments, *134-135*
Installation, *130-131*
Turning, *129*
Paneling, *12*
Cap rails, *38*
Ceilings, *52-55*
Frame-and-panel wainscoting, *39*, *40*, *41*, *46*
cope-and-stick frames, *47-48*
installation, *49-51*
raising the panels, *48-49*
Tongue-and-groove, 39, *40*, *42-43*
installation, *44-45*
Panels:
Raised, *46*, *93*
Patching compounds, *back endpaper*
Picture-frame casings, *58*, *61-65*
Picture rails, *12*, 30
Portable electric planers, *88*
Power miter saws, *15*, *68*

R-S-T-U

Railings. *See* Balusters; Handrails; Newel posts
Ripping, *16*
Rosettes, *58*, *73-74*
Shop-made rosette cutters (Shop Tip), *74*
Routers:
Mortising jigs, *75*
Tables, *14*
Router tables, *14*
Schuttner, Scott, *10*-11
Scribing, *front endpaper*, *42*
Shapers, *14*
Shop Tips:
Moldings, *29*
Windows, *68*, *72*, *74*, *80*
Sliding compound miter saws, *15*, *16*, *18*
Stairs, 11, *12-13*, *114*-115
Anatomy, *116-117*
Rise-and-run calculation, *115*, *117*
Risers, *125-126*
Stringers
installation, *122*
length calculation, *117*
making, *118-121*
Treads
balusters, *127*
installation, *126*
making, *123-125*
See also Balusters; Handrails; Newel posts
Stool-and-apron casings, *57*, *58*, *69-72*
Table saws, *14*, *16-17*
Taylor, Grant, *6-7*
Tongue-and-groove wainscoting, 39, *40*, *42-45*
Tools:
Carpenter's squares, *front endpaper*
Combination planes, *25*
Coping saws, *15*
Finish nailers, *14*, *21*
Log-builder's scribes, *42*
Miter boxes, *15*
Molder/planers, *15*, *20*
Portable electric planers, *88*
Power miter saws, *15*, *68*
Routers
mortising jigs, *75*
tables, *14*
Shapers, *14*
Sliding compound miter saws, *15*, *16*, *18*
Table saws, *14*, *16-17*
Vertical vises, *88*

V-W-X-Y-Z

Vertical vises, *88*
Wainscoting. *See* Paneling
Windows, *12*, 57
Casings, *12*
butted sill, *65*
picture-frame, *58*, *61-65*
stool-and-apron, *57*, *58*, *69-72*
temporary brace to hold window aprons (Shop Tip), *72*
Double-hung, *59-60*
Glass-stop moldings, *56*, *79-80*
installing the molding with a hammer (Shop Tip), *80*
Glazing bar half-lap joints, *81-83*
Jambs, *61-62*
Rosettes, *58*, *73-74*
shop-made rosette cutters (Shop Tip), *74*
Sashes, *75-79*
Wood, 19
Patching compounds, *back endpaper*

ACKNOWLEDGMENTS

The editors wish to thank the following:

FINISH CARPENTRY BASICS
Delta International Machinery/Porter Cable, Guelph, Ont.; Hitachi Power Tools U.S.A. Ltd., Norcross, GA; Jet Equipment and Tools, Auburn, WA; Tool Trend Ltd., Concord, Ont.; Williams and Hussey Machine Co., Inc., Wilton, NH

MOLDING
Adjustable Clamp Co., Chicago, IL; Delta International Machinery/Porter Cable, Guelph, Ont.; Great Neck Saw Mfrs. Inc. (Buck Bros. Division), Millbury, MA; Jet Equipment and Tools, Auburn, WA; Lee Valley Tools Ltd., Ottawa, Ont.; Blair McDougall, Brome Lake, Que.; Richards Engineering Co., Ltd., Vancouver, BC; Sandvik Saws and Tools Co., Scranton, PA; Sears, Roebuck and Co., Chicago, IL; Stanley Tools, Division of the Stanley Works, New Britain, CT; Tool Trend Ltd., Concord, Ont.; Walter Tomalty Enterprises Ltd., Montreal, Que; Williams and Hussey Machine Co., Inc., Wilton, NH

PANELING
Adjustable Clamp Co., Chicago, IL; Delta International Machinery/Porter Cable, Guelph, Ont.; Hitachi Power Tools U.S.A. Ltd., Norcross, GA; Jet Equipment and Tools, Auburn, WA; Lee Valley Tools Ltd., Ottawa, Ont.; Richards Engineering Co., Ltd., Vancouver, BC; Shopsmith, Inc., Montreal, Que.; Stanley Tools, Division of the Stanley Works, New Britain, CT; Tool Trend Ltd., Concord, Ont.; Williams and Hussey Machine Co., Inc., Wilton, NH

DOORS
Adjustable Clamp Co., Chicago, IL; American Tool Cos., Lincoln, NE; Delta International Machinery/Porter Cable, Guelph, Ont.; De-Sta-Co, Troy, MI/Wainbee Ltd., Montreal, Que.; General Tools Manufacturing Co., Inc., New York, NY; Great Neck Saw Mfrs. Inc. (Buck Bros. Division), Millbury, MA; Hitachi Power Tools U.S.A. Ltd., Norcross, GA; Jet Equipment and Tools, Auburn, WA; Lee Valley Tools Ltd., Ottawa, Ont.; Putnam Products, Old Saybrook, CT; Record Tools Inc., Pickering, Ont.; Sears, Roebuck and Co., Chicago, IL; Stanley Tools, Division of the Stanley Works, New Britain, CT; Tool Trend Ltd., Concord, Ont.

WINDOWS
Adjustable Clamp Co., Chicago, IL; American Tool Cos., Lincoln, NE; Delta International Machinery/Porter Cable, Guelph, Ont.; De-Sta-Co, Troy, MI/Wainbee Ltd., Montreal, Que.; General Tools Manufacturing Co., Inc., New York, NY; Great Neck Saw Mfrs. Inc. (Buck Bros. Division), Millbury, MA; Jet Equipment and Tools, Auburn, WA; Marvin Windows and Doors Inc., Toronto, Ont.; Blair McDougall, Brome Lake, Que.; Record Tools Inc., Pickering, Ont.; Robert Sorby Ltd., Sheffield, U.K./Busy Bee Machine Tools, Concord, Ont.; Sears, Roebuck and Co., Chicago, IL; Stanley Tools, Division of the Stanley Works, New Britain, CT; Tool Trend Ltd., Concord, Ont.; Vermont American Corp., Lincolnton, NC and Louisville, KY

STAIRS
Adjustable Clamp Co., Chicago, IL; American Tool Cos., Lincoln, NE; Colonial Elegance Inc., Montreal, Que.; Delta International Machinery/Porter Cable, Guelph, Ont.; Freud Westmore Tools, Ltd., Mississauga, Ont.; Great Neck Saw Mfrs. Inc. (Buck Bros. Division), Millbury, MA; Lee Valley Tools Ltd., Ottawa, Ont.; Record Tools Inc., Pickering, Ont.; Sandvik Saws and Tools Co., Scranton, PA; Sears, Roebuck and Co., Chicago, IL; Skil Canada, Ltd., Toronto, Ont.; Stanley Tools, Division of the Stanley Works, New Britain, CT; Thoroughbred Sawhorses and Equipment, Division of the Ivy Group, Inc., Valparaiso, IN; Tool Trend Ltd., Concord, Ont.; Vermont American Corp., Lincolnton, NC and Louisville, KY

The following persons also assisted in the preparation of this book:

Lorraine Doré, Graphor Consultation, Geneviève Monette

PICTURE CREDITS

Cover Robert Chartier
6,7 Alan Briere
8,9 Marie Louise Deruaz
10,11 Charles Mason
30 Courtesy Ornamental Mouldings
39 Courtesy Patella Industries, Inc.
114 Courtesy Boiseries Raymond, Inc.
128 Courtesy Boiseries Raymond, Inc.
132 Charles Mason